Orange Pilled

A Brief Introduction to Bitcoin

Daniel Howell, Ph.D.

Blue Ridge Books

Published Block 926550

Orange Pilled: A Brief Introduction to Bitcoin

ISBN 979-8-218-87353-0

Daniel Howell, Ph.D.

Copyright 2025. All rights reserved.

Send inquiries to

Blue Ridge Books, LLC

PO Box 4652 Lynchburg, VA 24502

Contents

Scriptures Addressing Money

You shall not steal.
-Exodus 20:15

Do not steal. Do not lie. Do not deceive one another.
-Leviticus 19:11

The rich rule over the poor, and the borrower is slave to the lender.
-Proverbs 22:7

Do not use dishonest standards when measuring length, weight, or quantity. Use honest scales and honest weights. I am the Lord your God, who brought you out of Egypt.
-Leviticus 19:35-36

Do not have two different weights in your bag, one heavy and one light. Do not have two different measures in your house, one large and one small. You must have accurate and honest weights and measures so that you may live long in the land the Lord your God is giving you. For the Lord your God detests anyone who does these things, anyone who deals dishonestly.
-Deuteronomy 25:13-16

A good man leaves an inheritance for his children's children, but a sinner's wealth is stored up for the righteous.
-Proverbs 13:22

Honest scales and balances belong to the Lord. All the weights in the bag are of his making.
-Proverbs 16:11

If you lend money, you shall not be like a moneylender (banker) to him, and you shall not exact interest from him.
-Exodus 22:25

The Lord detests differing weights, and dishonest scales do not please him.
-Proverbs 20:23

Be not one of those who gives pledges, who puts up securities for debts.
-Proverbs 22:26

You are to use accurate scales, an accurate ephah, and an accurate bath.
-Ezekiel 45:10

Anyone who has been stealing must steal no longer, but must work, doing something useful with their hands, that they may have something to share with those in need.
-Ephesians 4:28

Hear this, you who trample the needy and do away with the poor of the land, saying, "When will the New Moon be over that we may sell grain, and the Sabbath be ended that we may market wheat?" – skimping on the measure, boosting the price and cheating with dishonest scales.
-Amos 8:4-5

Am I still to forget your ill-gotten treasures, you wicked house, and the short ephah which is accursed? Shall I acquit someone with dishonest scales, with a bag of false weights? Your rich people are violent; your inhabitants are liars, and their tongues speak deceitfully.
-Micah 6:10-12

The Lord has a charge to bring against Judah; he will punish Jacob according to his ways and repay him according to his deeds… The merchant uses dishonest scales and loves to defraud.
-Hosea 12:2;7

The Lord detests dishonest scales, but accurate weights find favor with him.
-Proverbs 11:1

Your silver has become dross; your choice wine is diluted with water.
-Isaiah 1:22

Then Peter said, "Ananias, how is that Satan has so filled your heart that you have lied to the Holy Spirit and kept some of the money you received for the land? Didn't it belong to you before it was sold? And after it was sold wasn't the money at your disposal?
-Acts 5:3-4

And forgive us our debts, as we forgive our debtors.
-Matthew 6:22

INTRODUCTION

GETTING PILLED

"Remember... All I'm offering is
the truth. Nothing more."

Morpheus, *The Matrix*

The "pill" metaphor originates from the 1999 film *The Matrix,* where taking the red pill represents choosing to see harsh reality over comforting illusion. In online culture, particularly since the mid-2010s, different colored "pills" have emerged as shorthand for major paradigm shifts or awakenings about society, power, relationships, economics, or existence itself. The best-known are the red pill and black pill. The red pill refers to recognizing biases in society, especially in politics and media, that are driven by a progressive ideology that seeks to disrupt, or even remake, traditional family, economic, and political systems. It leads to a rejection of the mainstream, "woke" political agenda and progressive narratives on race, gender, immigration, or institutional trust.

The black pill is its nihilistic cousin: it accepts the same premises but includes disillusioning possibilities about the sexual marketplace and genetic determinism (looks, height, race, etc.). It concludes that if you lost the cultural lottery, no amount of effort, money, or "game" will matter; romantic success is essentially preordained, society is brutal and dysgenic, and coping or withdrawal is the only rational response. Black-pilled communities tend to be deeply fatalistic, sometimes veering into extremist ideology.

Less common but still recognized are the white pill and the blue pill. The white pill is a deliberate turn toward hope, faith, or accelerationism (believing things must get worse before they get better). The white pilled believe that technological and/or spiritual breakthroughs are coming, so hang in there. The blue pill refers to remaining asleep, choosing to ignore the other pills of enlightenment and to remain blissfully ignorant inside the mainstream narrative. In the Bitcoin community, these people are often referred to as *normies.* While this word can be taken as a pejorative, it usually should not be interpreted that way. In most cases, it simply refers to a person who has not yet seen the light.

Being orange pilled specifically refers to the moment someone grasps the revolutionary implications of Bitcoin and sound, decentralized money. It is the realization that fiat currency is a hidden tax enforced by central banks and governments through inflation and monetary debasement, that most of what people think of as "money" is just debt-based paper tokens controlled by a cartel, and that Bitcoin represents a peaceful but unstoppable opt-out from that system. Orange-pilled individuals typically shift from regarding Bitcoin as speculative internet gambling to viewing it as the hardest, most verifiable form of property ever created; a permissionless form of money that cannot be confiscated, frozen, or even taxed without *your* permission. The term borrows the color of Bitcoin's logo and has become the dominant "pill" metaphor inside cryptocurrency circles, spawning phrases like *"have you tried orange pilling your normie friends?"* It often leads to broader libertarian or Austrian-economic views and a deep skepticism of any financial asset or system that can be manipulated by a handful of people or printed at will.

So, being "pilled" is an awakening, an opening of your eyes to a reality you've not seen before and didn't even know existed but once seen cannot be unseen. My hope in writing this little book is that you will be awakened to the innerworkings of our monetary system, be appalled by what you see, and embrace the solution that is as simple as downloading an app.

CHAPTER 1

WHAT IS MONEY?

"It is well enough that people of the nation do not understand our banking and money system, for if they did, I believe there would be a revolution before tomorrow morning."

Henry Ford

Scarcity and Value

Have you ever stopped to ask, *"What is money?"* I'm ashamed to say I was 50 years old before I asked the question. To be fair, I was preoccupied with other things and the money we use seemed to be working, and we are never urged to question it. So, I found myself sleepwalking through life when it came to money. Oh, forces outside the Matrix did try to wake me. In 2008, the Great Financial Crisis nudged me and attempted to stir my consciousness, but I rolled over and went back to sleep. Then COVID arrived in 2020, and all the chaos that came with it. Stimulus checks in the mail finally did it; they jolted me from my slumber, and I finally asked the question: What is money?

So, why are these paper slips we call dollars money? Or euros? Or pesos? They say money doesn't grow on trees, but why *don't* we use leaves as money? Or gravel?

We don't use leaves or gravel as money because leaves decompose, and both leaves and gravel are extremely abundant. In order to be good money, the object used as money must be *durable* and *scarce*. The scarcer something is, the more valuable it is. Leaves are neither durable nor scarce. Gravel is durable, but it is not scarce.

For centuries, gold served as money because it is both durable and scarce. However, gold is still being mined, and new gold is added to the existing supply by 2-4% each year. This means that gold is slowly becoming less scarce. In addition, gold is difficult to verify. Gold bars and coins must be melted down, x-rayed, or subjected to other expensive treatments to confirm they are genuine.

Other things have been used as money. Many years ago, both salt and seashells were used as money. Back then, both were hard to obtain, but today they are too easy to acquire. Due to its past use as payment, we get the word "salary" from the Latin word for salt (*sal*).

Good Money

Leaves and gravel have never been used as money because they are too abundant. Salt and shells were reasonable for a time, but ultimately, they were too abundant, as well. The reason gold won the "money wars" is because gold is scarce, and so it is best at storing value over time. Storing value into the future is one of the main functions of money, but there are two other important functions as well: money is a medium of exchange and a unit of account. Gold can do all these things reasonably well.

Figure 1: What is money?

Let's examine the three functions of money in more detail. They are 1) a *unit of account*, 2) a *medium of exchange*, and 3) a *store of value*.

1. Unit of Account.
'Unit of account' means we price things in our chosen monetary unit. Roughly, a candy bar is worth $2, a cell phone is worth $500, and a car is worth $25,000. It is usually difficult to price things in more than one monetary unit and so only one form of money is typically used by a coherent society.

2. Medium of Exchange. How much money do you want? If you really think about it, for most people the answer is zero. Good money has no other value than what you can buy with it.[1] So, you don't want money, you want what you can *obtain* with money (free time, vacations, material possessions, and even power and influence). Dollars are just paper and ink, but dollars allow you to buy the things you want or need. It is something you can *exchange* for things you want or need. Without money, we are forced to barter. Suppose you own an apple orchard and need a new car. You could trade apples for the car, but the car seller may not want apples. And how many apples would it take to equal a car? Using a medium of exchange (i.e., money) avoids the problems associated with barter.

3. Store of Value (SoV)
Good money stores value into the future. In 1850, one ounce of gold (worth $18) would buy you a new, high-end suit. Today, one ounce of gold ($2,500 in 2024) will buy you a new, high-end suit. The cost of a suit has changed a lot in terms of dollars, but not in terms of gold. You can see that gold has been a better *store of value* (SoV) than dollars. The reason, of course, is because dollars can be printed at will; gold must be mined from the Earth.

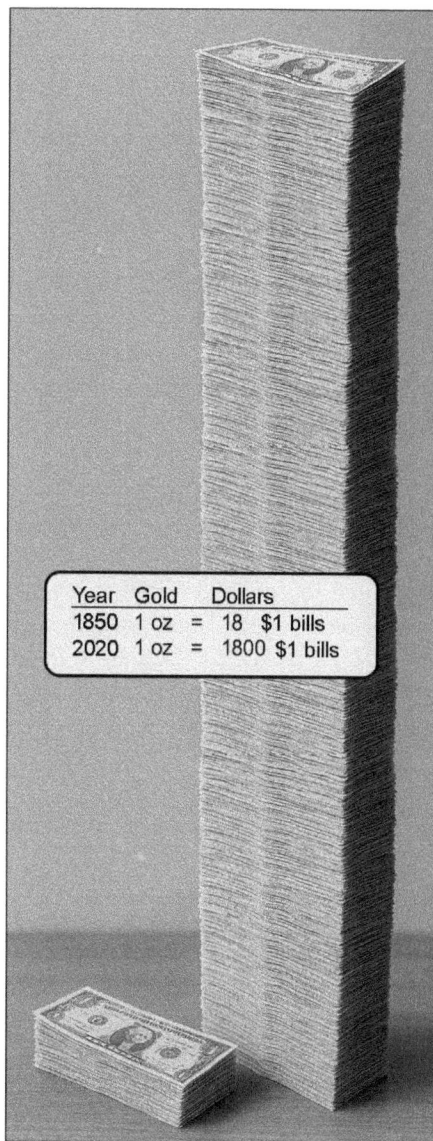

Year	Gold		Dollars
1850	1 oz	=	18 $1 bills
2020	1 oz	=	1800 $1 bills

Figure 2: The Price of 1 oz Gold

---

[1] Good money has only one use-case: being money. Additional use cases make the object less valuable as money. The fact that gold is used in jewelry and electronics actually makes gold less valuable as a money.

3

Salability

1. Salable across scale means that a money can be easily divided or grouped to make both small and large purchases. The US dollar has remarkably low salability of scale: the largest unit ($100 bill) is only 10^4 times larger than the smallest unit (one penny). The problem is compounded by the fact that the smallest units are becoming worthless. The only reason this lack of salability is not obvious is because we usually perform transactions electronically or by check. Consider using cash (i.e., physical paper bills) to buy a new car or house. It would require a suitcase of $100 bills.

2. Salability across time means your money holds its value into the future. One ounce of gold would buy you a nice suit in 1850, 1900, 1950, and today. So, if you put one ounce of gold in a vault and save it for 50 years, it will buy the same amount then as it does now. Gold is a reasonably good store of value over time, but it's not perfect. New gold is added to the supply every year. In fact, the rate of gold production is accelerating due to advances in gold mining (Figure 3). This will surely impact gold's ability to be a good store of value into the future. The reason seashells failed as money is because technological advances made it too easy to obtain shells from the ocean floor.

3. Salability across space means that the chosen money is easy to transport. Here, the dollar beats gold hands down. Gold is heavy and therefore difficult to carry on your person in significant amounts. It is also dangerous to carry over long distances; piracy on the seas and bandits chasing down trains were major problems during the Golden Age. Paper money is much lighter, and since paper money can be printed, stolen money can be replaced when insured. Indeed, the difficulty of carrying gold is the reason why paper money exists. People stored gold in banks and exchanged the paper receipts instead.

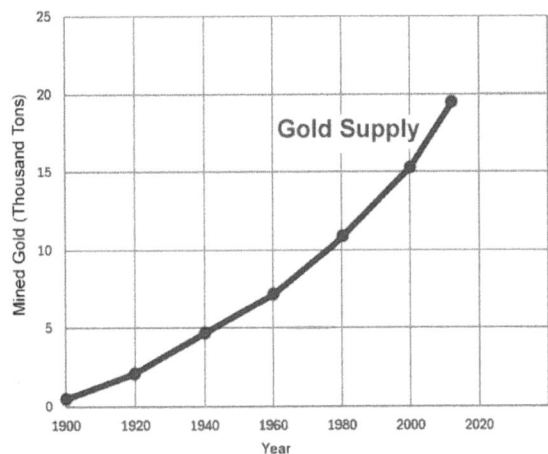

Figure 3: Gold Production Increasing

Central Banks & Paper Money

Salt, seashells, and gold have all been tried as money and ultimately failed for one reason or another. Today, we use paper as money.[2] Paper created by the Federal Reserve (the "Fed").

The Federal Reserve is the central bank of the United States. It is a coalition of private banks; thus, it is not part of the Federal government. It also has no reserves, so the name *Federal Reserve* is an apparent deception. As a coalition of private entities with monopoly

_____

[2] I guess money really does grow on trees.

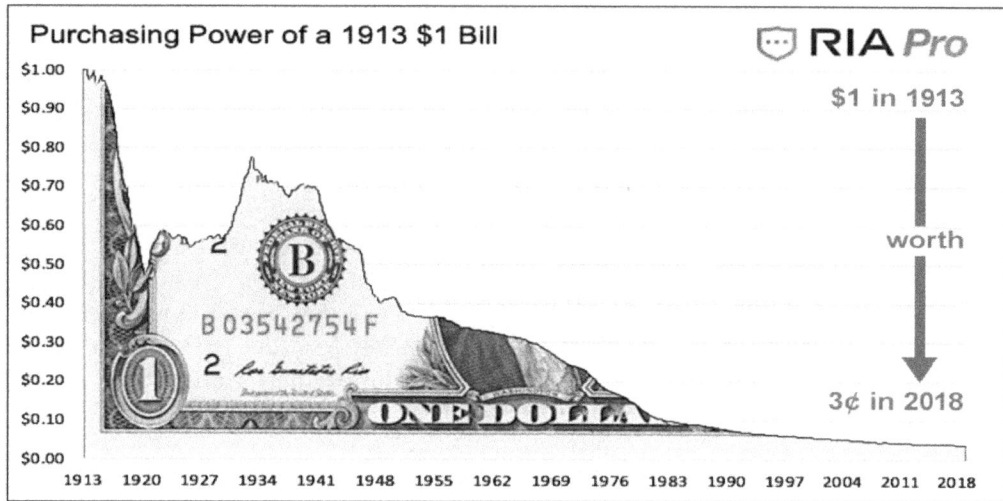

Data Courtesy: St. Louis Federal Reserve

Figure 4: Debasement of the Dollar by the Fed

control over the issuance of money, it is quite literally a cartel. A cartel that, through the coercive power of government, requires us to use *their* money under *their* rules.

The Fed was created by an act of Congress in 1913. (Coincidently the same year the income tax was created). The United States did not have a central bank prior to that except for two brief periods in the 1800's. Thomas Jefferson believed that central banks are more dangerous to freedom than standing armies. He is being proved right.

One mandate of the Federal Reserve central bank is to preserve the value of the dollar. It has obviously failed. The purchasing power of a 1913 one-dollar bill is now just three pennies, and pennies are so worthless they just stopped making them.[3] Astoundingly, the dollar has lost over 95% of its purchasing power since the Fed was created (Figure 4). This devaluation is reflected in higher prices for goods and services. In other words, infla-tion.

Inflation happens when you print money and thus *inflate* the supply. It results in a loss of purchasing power and shows up as rising prices for everything. Your grandfather could snag everyday essentials at a Five & Dime store for mere pennies; today, even the Dollar Store rarely delivers on its name; most items on the shelf are $5 - $10. Back in 1850, you could buy an ounce of gold for just $18. By 2020, that same ounce commanded 18 *hundred* dollars. By 2025, just five years later, the price soared past $4,000. To be clear, it's not the value of gold that is rising, it's the value of the dollar that's dropping. Fast. Like in freefall.

---

[3] RIP Penny 1793 – 2025.

Gold-backed Money

Although the value of the dollar is dropping quickly now, that wasn't always the case. Until 1971, the dollar was remarkably stable. The price of a can of Campbell's tomato soup held steady at $0.10 from 1895 to 1970. Your great grandfather paid 10 cents in 1920, and your grandfather paid 10 cents in 1960. But in 1971 the unraveling began. Today, you pay $1.29 for that same can of soup (Figure 5).

The value of the dollar held steady for many years because it was pegged to gold. When the Federal Reserve was created, dollars were fixed at $18 per ounce of gold. Of course, the government couldn't resist the urge to spend more money than it collected in taxes, and the Fed was happy to print the difference. When they could no longer hide this theft, Franklin D. Roosevelt confiscated (i.e., stole) every American's gold via an executive order (EO 6102). The gold was collected and placed into Fort Knox for "safekeeping."

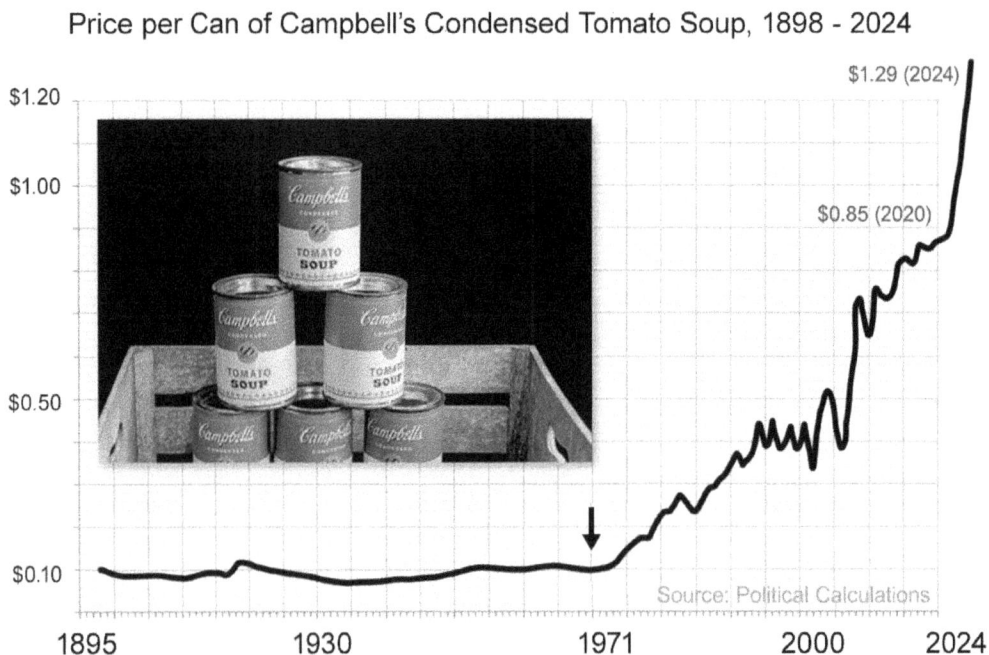

Figure 5: The Price of Campbell's Tomato Soup

If the dollar lost 30x its purchasing power, why has soup only gone up 15x in price? Because the production cost of soup has *decreased* 2x due to technological advancements. Without inflation created by the Federal Reserve, the price of soup would have actually *dropped* from $0.10 to less than $0.05. In fact, without the Fed, the price of *everything* would go down over time, not up.

6

Shortly after, gold was repriced to $38 per ounce. In other words, the value of the dollar was cut in half almost overnight in 1933 by the stroke of a pen. You can see this dramatic revaluation of the dollar following EO 6102 in Figure 6.

During World War 2, our Europeans allies, threatened with the prospect of losing a war and being looted, decided to move their gold to Fort Knox. As the war came to a close, and since we had saved Europe and held everyone's gold in America, an international agreement at Bretton-Woods, NH established the US dollar as the global reserve currency in 1944, meaning all countries would conduct international trade in dollars. This created a global demand for dollars, and the dollar Ponzi scheme began in earnest. By printing dollars and exporting them to the world, Americans benefitted from money printing while disseminating the effects of inflation. That is, we got rich at the expense of the rest of the world. But how can the Fed just print dollars when they're supposed to be pegged to gold? By the 1960's, lots of national leaders were asking that question.

Fiat Money
When countries started to catch on to the printing scheme in the 1960's, several demanded their gold back. In 1968, the French even sent warships to New York Harbor to pick up their gold. Realizing the charade was over and fearing that other nations would demand their gold (which was likely already stolen), Richard Nixon de-pegged the dollar from gold on August 15, 1971. This event is called the Nixon Shock, and it officially converted the dollar into *fiat currency*. Going forward, no one would be able to exchange their dollars for gold. Nixon claimed this would be a temporary measure to "stabilize the dollar," but as Ronald Reagan would later point out, nothing is more permanent than a temporary government program.

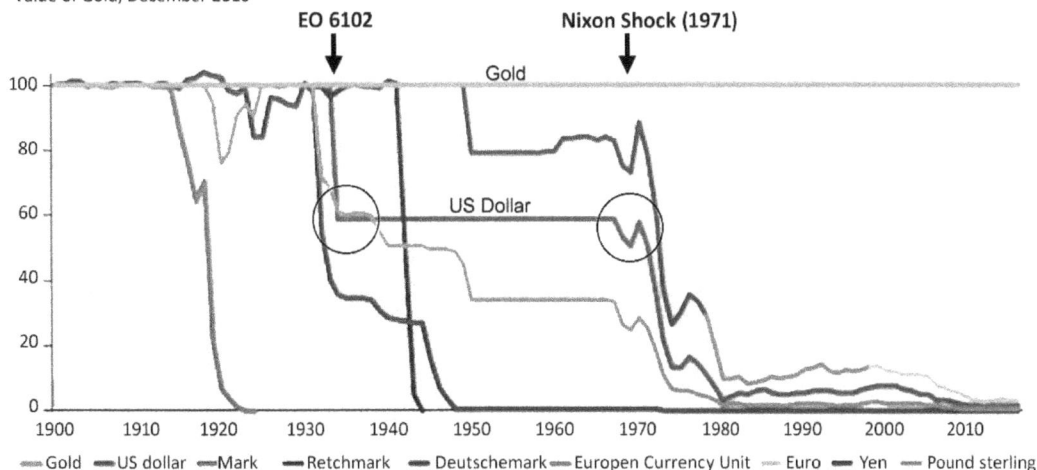

Figure 6: Collapse of Fiat Currencies

The temporary suspension of the convertibility of dollars into gold is still in effect today, over 50 years later. You can see its impact on the dollar in Figure 6.

Fiat money is backed by nothing but the "full faith and credit" of the issuer. Once the dollar was de-pegged from gold, the government and the Fed quickly proved why they deserve neither faith nor credit from the People; the money printer immediately went brrrr and the cost of a soup climbed from 10 cents to a buck thirty, a nearly 15-fold increase in price. The price of soup (and everything else) has tracked the money supply almost perfectly (Figure 7). Trillions of dollars have been printed since 1971, and our national debt has soared with it. Today, our debt is over $37 trillion and rising by roughly $1 trillion every 100 days. It took 39 presidents to reach our first $1 trillion in debt; the past 7 have added another $36 trillion (Figure 8).

For centuries, gold's purchasing power has stayed remarkably consistent, while dozens of fiat currencies have marched inexorably toward zero (Figure 6). The dollar is not at zero (yet) and it will likely be around for years to come, but all fiat currencies go to zero eventually. Will the dollar somehow be the exception to the rule? Based on the money printing of the past few years and knowing that our economy will collapse if the printer stops, my bet is the dollar will *not* be exempt from the laws of nature and economics. It will continue to be debased if only to pay off our debts.[4]

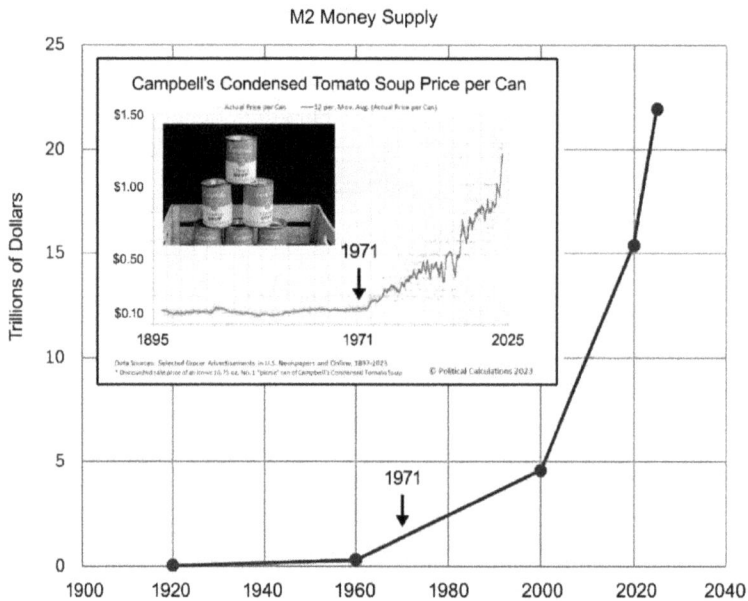

Figure 7: Money Supply and Inflation

---

[4] Stable coins like Tether (USDT) are digital tokens pegged to the dollar. Tether is pegged 1:1 to the dollar. These coins will perhaps breathe new life into the dying dollar and allow it to function for another decade or two. It does this by exporting inflation to a new cohort of victims. While nations around the globe de-dollarize, the dollar is still prized by individuals around the world because it is

I said the gold in Fort Knox was likely already stolen by 1971. Why do I and many others believe the gold was looted by bankers, politicians, and other elites? Because Fort Knox has never been legitimately audited (there was a partial audit in 1970), and even President Trump has been unable to achieve an audit. Until proven otherwise, we can assume there is no gold in Fort Knox.[5]

The devasting impact of uncoupling the dollar from gold goes far beyond a pricy bowl of soup. Money is the base layer of civilization. Much of the societal decay we are witnessing can be traced back to the Nixon Shock and the subsequent debasement of our money.

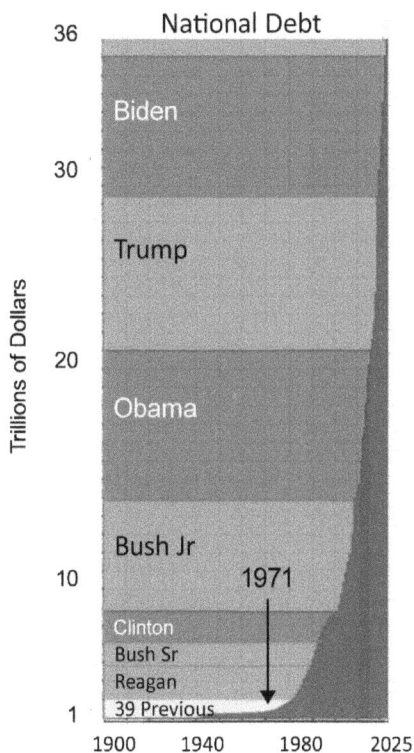

Figure 8: Debt by Year & POTUS

---

better at storing value than their local fiat currencies. Billions of these people, especially in Latin America, are unbanked. By simply downloading a stable coin app they can have access to dollars and join the global economy, even thwarting their governments' capital controls. To expand into these markets, Tether is rapidly minting USDT tokens, and since each USDT is pegged to a dollar, Tether Inc. is becoming the largest buyer of US treasury bonds. The US government is catching on and now sees Tether as a new way to extend the dollar's life.

[5] Curiously, large shipments of gold have been sneaking into the United States from other countries recently. In the 19th century, it was common for banks to move gold from one bank to another to deceive people into thinking all banks had full gold reserves. Is the same thing happening now?

CHAPTER 2

WHAT'S WRONG WITH OUR MONEY?

"I believe that banking institutions are more dangerous to our liberties than standing armies... If the American people ever allow private banks to control the issue of their currency, the banks and corporations that will grow up around them will deprive them of all property until their children wake up homeless on the continent their fathers conquered."

Thomas Jefferson

Broken Money = A Broken Society

From just our brief review of the dollar in the previous pages, you can tell there are real dangers associated with using fiat money. In addition to a run-away national debt and rising price of soup, printing money has destroyed the middle class and widened the gap between the have-nots and the have-yachts. The poor, who have no assets and save in dollars (if they can save at all), are getting poorer every year. The rich put their wealth into inflation-beating assets like real estate and stocks. And money is cheaper for rich people; they get loans at much lower rates than the average person. In 2024, the typical mortgage rate was 6.5-7%, yet Blackrock bought up single-family homes with mortgage rates less than 1%. This favoritism for the rich and those closest to the money printer is called the *Cantillon Effect*, and it results in the rich getting richer while the poor get poorer. We were told it would trickle down, but it never did. At its core, fiat money is a debt-based, theft-based system that favors the rich at the expense of the middle and lower classes.

The US dollar is not the first fiat currency. Many nations have tried it, and all have failed. The average lifespan of a fiat currency is just 20-30 years, yet the US dollar has been fiat for more than 50 years. Its unusual longevity is explained by its unique status as the world reserve currency. We have delayed the effects of fiat money printing by exporting those effects to the world at large. But the gig is up. Inflation is rising and many nations are reconsidering the reserve status of the dollar. In October 2024, the BRICS[1] nations met to discuss de-dollarization. If the rest of the world abandons the dollar, the dollar will collapse almost overnight. Indeed, most of the wars of the past 50 years have been to protect the dollar, not to fight terrorism or some other gaslighting nonsense. Sadam Hussein never had weapons of mass destruction, but he did have plans to sell oil *not* in dollars. Animosity towards the US dollar by nation-states only increased when the Biden administration weaponized the dollar for political ends. Russia, like most countries, holds dollars in reserve as a national asset. The Biden administration first froze the movement of those dollars internationally, then wholly confiscated them. To add insult to injury, they gave the confiscated dollars to Russia's enemy: Ukraine.

Seizures and confiscations are overt theft, but banks have other ways of stealing from nations that are more clandestine. There is a global coalition of banks called the International Monetary Fund (IMF). The IMF purportedly exists to help third world countries develop into first world nations. It does this by providing loans, but unsurprisingly the loans come with strings attached. The loans are designed to end in default, and when that inevitably happens the resources of the country (e.g., copper mines, ports, or oil fields) are possessed as repayment. In its 50-year history, the IMF has not lifted a single nation out of poverty – not one. But it has siphoned 64 TRILLION DOLLARS from third world countries into the coffers of banks and the corporations around them. Such predatory lending is not simply immoral, it is evil.

Although predatory lending is covert, there is another form of theft still more secretive: inflation. Legalized money printing (i.e., legal counterfeiting) by a monopoly bank does not make counterfeiting moral or magically abolish the consequences. And unfortunately,

---

[1] Brazil, Russia, India, China, South Africa, plus Egypt, Ethiopia, Iran, Indonesia, and UAE.

money creation doesn't just happen at the central bank, it happens at every bank, through fractional reserve lending.

How Did We Get Here?

How did we even get banks and paper money? Recall from chapter 1 that we need something both durable and scarce to have good money. Ideally, money should be a commodity (or at least fixed to a commodity), something not controlled by any single person or group. Oil, corn, and gold are commodities; they are products of nature that anyone can drill, grow, or mine to acquire. No one holds the patent on oil, or the copyright for corn. Of these three, gold makes the best money because it is scarce and durable.

When you hold gold as money, you can physically possess it. That is, it is a *bearer instrument*. When you transact with gold, you exchange your gold for a good or service you desire and thus transfer ownership to the seller. When the buyer and seller go their separate ways, the seller now has possession of the gold. The transaction was both instantaneous and final. So, gold is a bearer instrument that permits *instantaneous final settlement* during transactions. So far so good, but remember that gold is cumbersome and dangerous to carry.

Because gold is cumbersome and dangerous to carry, banks evolved to keep gold for their customers and provide them with written receipts. Traveling with paper receipts is both easier and safer. Buyers and sellers began trading receipts knowing they could be exchanged for real gold at any time at the bank. Like gold, cash is a bearer instrument that permits instantaneous final settlement.

Making Money (Literally)

This system worked well until the bankers got greedy. They quickly realized they could write more receipts than they actually held as gold in the vault. They wrote these extra receipts and gave them out as loans. This slight-of-hand is called *fractional reserve lending*. It silently debases the value of everyone's receipts (i.e., money). If done in small amounts, the debasement is unnoticeable. In large amounts, it shows up as inflation. Since their greed is insatiable, keeping it to small amounts is always impossible. When inflation gets out of control, people lose confidence in the paper receipts and the currency collapses.

The amount of dollars in circulation is not determined merely by printing paper at the central bank. Fractional reserve lending happens at every bank in America, made especially easy by computers. Do you think your local bank reaches into their vaults to loan you money for a car? Not so. Every time you take out a car loan or secure a mortgage or a line of credit, that money is created out of thin air with a few clicks of a keyboard. Multiply this trick by thousands of banks across America and you can see how this becomes a significant source of new money.

Let me say this again with names. Imagine you want to buy a car from your friend Adam for $25,000. Your other friend Bob prints $25,000 for you to give to Adam, but now you must pay Bob $25,000 plus $2,000 in interest. Bob created the money out of thin air in a matter of seconds, but you have to work over 1,000 hours (at $25/hour) to pay back the money you owed Bob. How on Earth is such a scheme even legal? As Jack Mallers, CEO of Strike, says: no one should have to work for money another man can print for free.

Throughout most of the past century, banks in the US were required to hold just $1 in the vault (called reserves) for every $9 they created and loaned out. During COVID, as the computers overheated with digital printing, the 10% reserve was eliminated. (Yes, reduced to zero). So, Banks literally create unlimited money out of thin air to lend out to you and me. We, however, must pay it back (with interest) from the sweat of our brows, slaving away at jobs most of us don't even like. Does this seem remotely fair to you? Is this how money is supposed to work?

Trusted Third Parties
Even if bankers had integrity (LOL) and eschewed fractional reserve lending, using cash (or gold) comes with a serious limitation: the buyer and seller must be physically together to transact. In today's world, most purchases are made using digital dollars over the internet, so this limitation is overcome. However, digital fiat comes with other problems (see Table 1). For one, it requires a trusted third party which may not be so trustworthy. Second, transactions are no longer anonymous, and every purchase can be tracked and controlled. Not surprisingly, governments love digital fiat currencies, and most governments are trying, openly and secretly, to create cashless societies. In Europe, cash transactions over €1000 are literally illegal. In America, some politicians wanted an accounting of every $600 we spend. Thankfully, they failed.

Third world countries might deprive their citizens of financial rights, but that would never happen in a developed nation like the USA or Canada, right?

In the deep freeze of winter in 2021, Canadian truckers gathered in Ottawa to protest vaccine mandates. The gathering turned into the largest protest in Canadian history. Despite the media propaganda saying otherwise, the protest was entirely peaceful. However, the government was displeased with the protest and got busy identifying as many protestors as possible. As they were identified, the government shut off their bank accounts. Hundreds of truckers were stuck in Ottawa without access to their money to buy food or even fuel to return home. Banking tyranny doesn't just happen in Venezuela or Argentina; it happened in Canada.[2] It happened in the US, too; even the Trump family was debanked.[3]

A Central Bank Digital Currency (CBDC) is digital fiat on steroids. Europe is rushing headlong into a CBDC. They expect to launch it in 2025 and have it fully operational by 2029. Agustin Carstens, former head of the Bank of International Settlements (BIS), openly confessed that CBDCs were desirable to banks because every transaction could be tracked.

Fiat money CBDCs are described in Revelation 13. There we read of an economic system controlled entirely by 'the beast.' In that system, no one is allowed to participate in the economy ("to buy or sell") unless they pledge their allegiance to this tyrant. This is fiat money at its worst; money that is tracked, controlled, and coercive.

---

[2]Notably, Bitcoiners around the world donated to a Bitcoin fund that was dispersed among the truckers. This momentous demonstration that Bitcoin can even defy governments was the event that orange pilled RFK Jr, a member of the Trump 47 administration.
[3]Being debanked orange pilled the Trump family, especially Eric Trump. The Trump family now owns an estimated 8,500 bitcoins.

Cash Payments	Third-party Payments
Must be done in person	Can be done remotely (online)
Immediate and final settlement	Final settle takes weeks or months
Requires no trust between parties	Requires trust in a third party
Requires trust in the money issuer (Fed)	Requires trust in the money issuer (Fed)
No fees	Costs additional fees (for third party)
Potentially dangerous	Much safer (from physical attack)
Payments cannot be tracked/blocked	Payments can be tracked/blocked
No need to share personal information	Must divulge lots of personal information

Table 1: Cash vs Third Party Payments

Theft & Debt

At the front of this book, I list over a dozen Scriptures that deal with money. I also refer to bible verses throughout the book. I do this because money is not merely an economic issue, it is a moral one. Here are four of my favorites verses on money:

> *"The Lord detests dishonest weights, but accurate scales are a delight to Him."*
> Proverbs 11:1

> *"A good man leaves an inheritance to his children's children."*
> Proverbs 13:22

> *"The borrower is slave to the lender."*
> Proverbs 22:7

> *"Thou shalt not steal."*
> Exodus 20:15

Our entire monetary system is built on debt. New dollars are washed into the system by lending, going first to those with assets or connections. The Fed simply creates ex nihilo trillions for the government to waste, leading to generational debt. Money that steals from the future is immoral. A monetary system designed to transfer wealth from the poor to the rich is immoral. A debt-based currency is immoral. Fractional reserve lending and rehypothecation (reusing someone else's money for your own purposes) are immoral. In times past, debasement was achieved by money-clipping or mixing metals, but those crude methods of theft pale in comparison to the modern monetary shenanigans of fiat money.

In 1970, my father bought a newly constructed brick house for $17,500. If my father had taken the advice of Proverbs 13:22 and set aside $18,000 for his grandson to buy a house in 2025, that money would not be enough even for a downpayment.[4] That scale of theft is unconscionable. In the 1960's, a man with a high school education could support a wife at

---

[4] Thankfully, my father passed down assets rather than mere cash.

home with children. By the 1990's, it was impossible to run a household on just one income, and kids were put into day-care. Today, it's nearly impossible to even get married and have a family at all. We've been told it's just natural for things to get more expensive over time. *That is a lie.* It's natural for things to get *cheaper* over time. Money printing, currency debasement, inflation, and the Cantillon effect are responsible for us getting poorer with every passing year. The World Economic Forum (WEF) proclaims that soon "you'll own nothing" but don't worry, "you'll be happy." They're getting richer and we're getting poorer by design. This is their plan for us. It's called Agenda 2030.

Figure 1: A post on X by the WEF

Fiat money steals from the working and productive members of society to enrich bankers, financiers, and wealthy investors, and as stated, this is an immoral transfer of wealth from the poor to the rich. Eventually, fiat money destroys completely the middle class, leading to masses of impoverished debt-slaves and a few wealthy overlords. As the rich get richer and the poor get poorer, it creates social unrest and political upheaval. It creates stress in families – arguments over money are the leading cause of divorce. It leads to more abortions – the cost of raising a child is a top reason for terminating pregnancies. Even art and architecture suffer under fiat money, as glorious cathedrals and magnificent bridges give way to drab cubes and tiny houses. As Bitcoin coder Jimmy Song puts it: *Fiat ruins everything.*[5]

The devastating effects of decoupling the dollar from gold can be seen in the following graphs[6] (in each graph, the arrow depicts 1971):

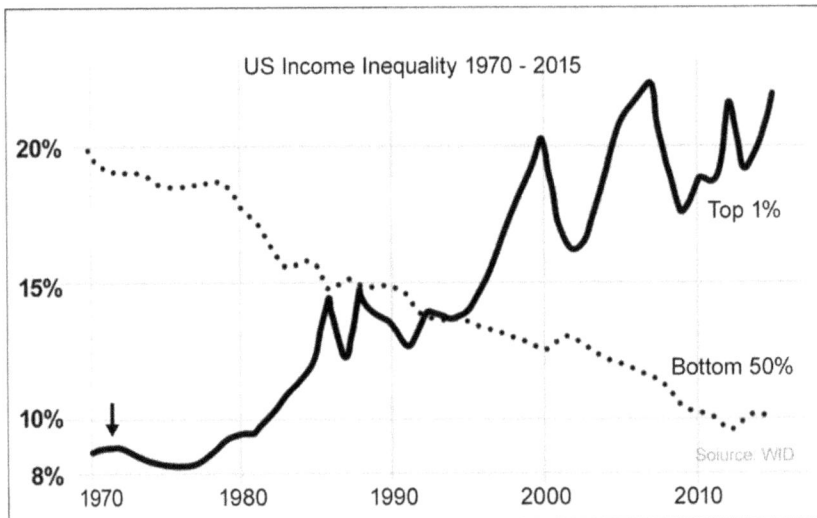

---

[5] *Fiat Ruins Everything*, by Jimmy Song. Available on Amazon.
[6] These graphs and many others can be viewed at www.wtfhappenedin1971.com

Growth in Productivity & Hourly Compensation, 1948 - 2017

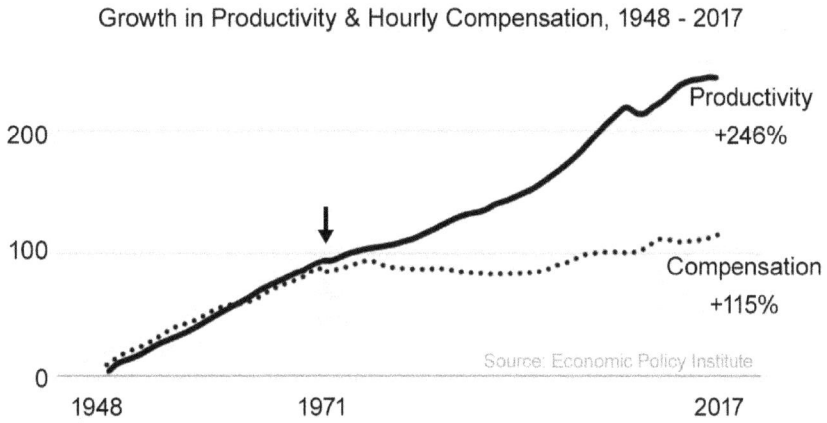

Income Growth, 1917 - 2012

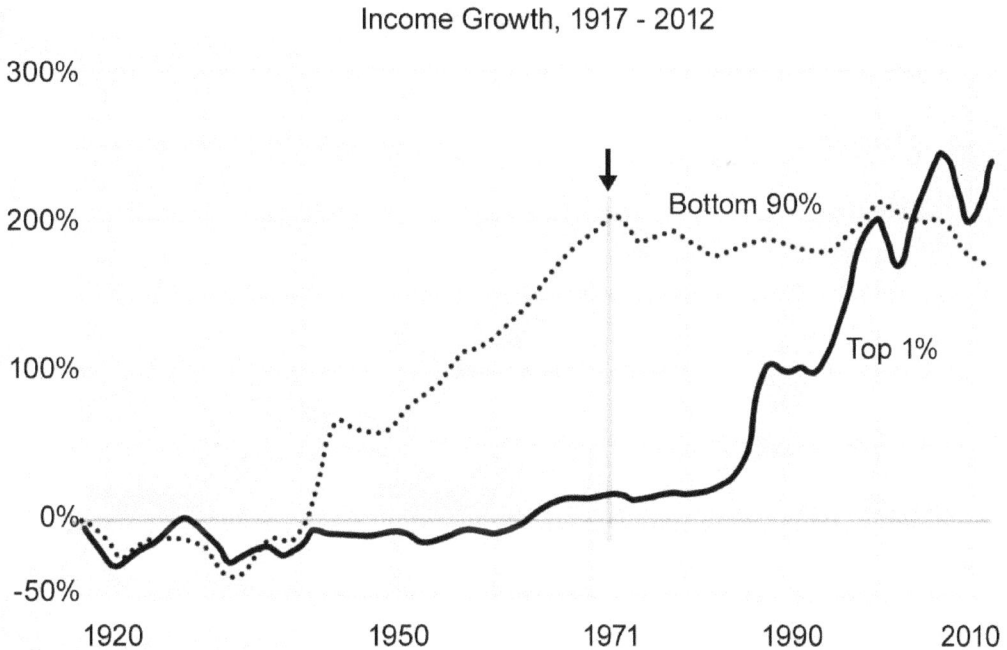

Average Black Income as % of Average White Income

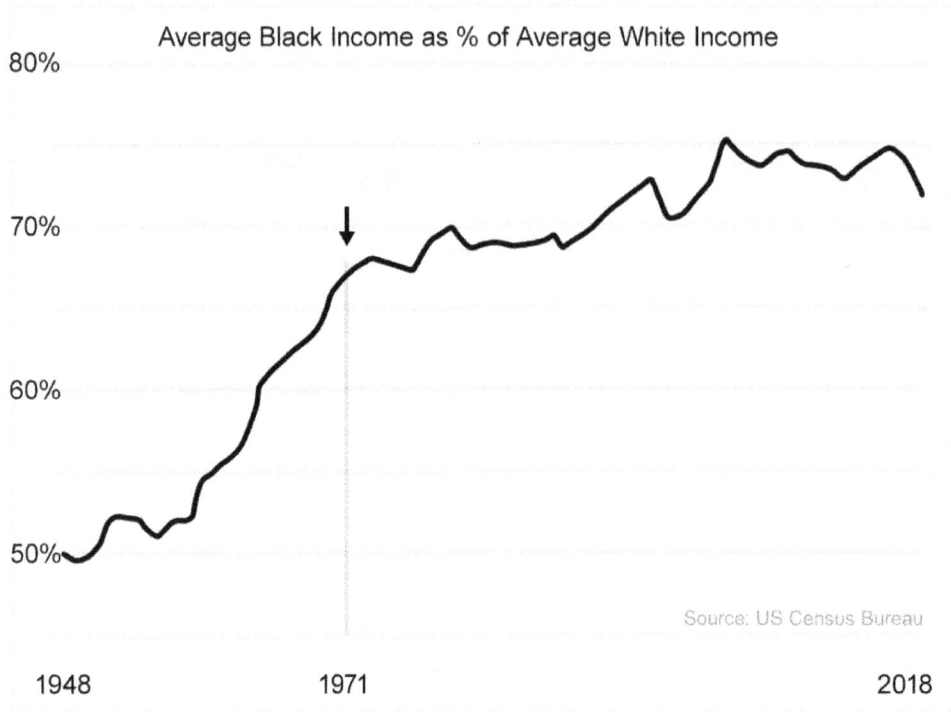

Source: US Census Bureau

Food and Fruit CPI

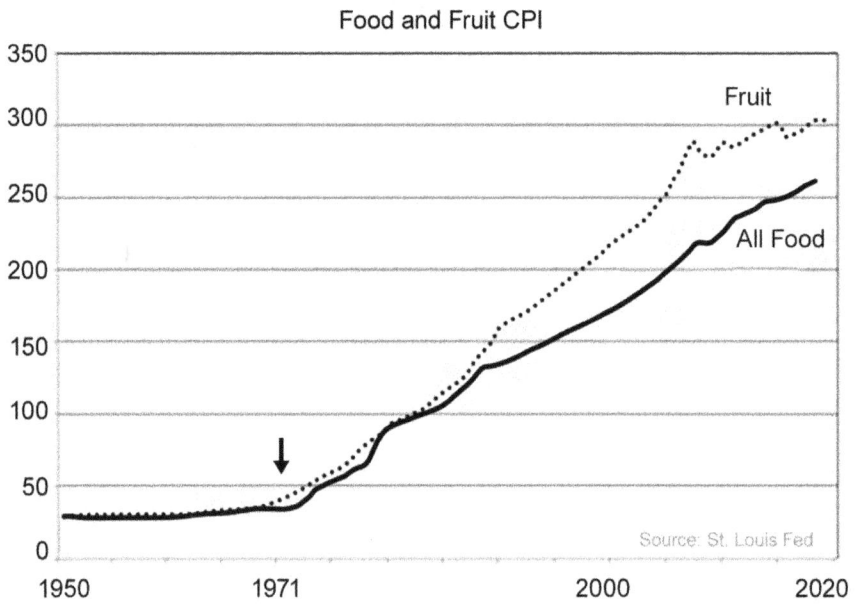

Source: St. Louis Fed

Societal Effects of Fiat Money

The preceding graphs provide just a snapshot of how fiat money can negatively impact society by debasing currency and transferring wealth from the poor to the rich. But these effects go beyond the financial: they foster widespread anxiety and despair, drug abuse, divorce, abortion, nihilism, high time preference[7], crime, and hopelessness. The elderly fear they don't have enough for retirement. The young feel the American dream is dead. They can't afford to even get married and start a family, so they cope with video games and worse. The whole nation seems to be under a curse. Our silver has become dross, and our choice wine is diluted with water (Isaiah 1:22). Maybe it's because our weights are dishonest and our scales are tilted (Proverbs 11:1).

It is remarkable to consider that dollars are literally monopoly money. They are paper slips printed by a central bank given legal monopoly power. I'm sure you are familiar with the board game called Monopoly. Imagine trying to win at Monopoly if the banker could just print more money whenever they ran out. Eventually, the banker would own everything, and you would own nothing. But you'll be happy, remember?

What Can We Do?

The dollar is in a precarious position. It is a fiat money that has been debased continually by a central bank inflating the money supply. It has been weaponized by our government, causing other nations to lose faith in it as the global reserve currency. History shows that when a fiat currency collapses, it happens quickly (in a matter of days). We could wake up one day soon to find the dollar collapsing. To maintain their control over money, governments will try (are trying) to implement CBDCs, which give them an even greater ability to track and coerce.

On the other hand, the dollar may languish on for decades more. After all, our politicians are good at kicking the can down the road. Even if the dollar continues for years to come, it is guaranteed to lose value. The dollar has lost 95% of its value in the first 50 years of my life. It is virtually guaranteed to lose at least another 95% should it continue for another 50 years.

So, what can we do? How can we protect our hard-earned wealth, and maybe right some wrongs in the process? Voting for the red team or the blue team doesn't seem to matter; our debt keeps rising higher and our culture keeps sinking lower.

Money is part of every purchase we make and every item we sell. Money is so pervasive in our lives and foundational to civilization that bad money can destroy a society, while good money can foster a healthy society. That's why sound money advocates believe that if we can *fix the money,* we can *fix the world.*

Enter Bitcoin.

---

[7] High time preference means living for the moment rather than planning for the future (low time preference)

19

Bitcoin: A Peer-to-Peer Electronic Cash System

Satoshi Nakamoto
satoshin@gmx.com
www.bitcoin.org

Abstract. A purely peer-to-peer version of electronic cash would allow online payments to be sent directly from one party to another without going through a financial institution. Digital signatures provide part of the solution, but the main benefits are lost if a trusted third party is still required to prevent double-spending. We propose a solution to the double-spending problem using a peer-to-peer network. The network timestamps transactions by hashing them into an ongoing chain of hash-based proof-of-work, forming a record that cannot be changed without redoing the proof-of-work. The longest chain not only serves as proof of the sequence of events witnessed, but proof that it came from the largest pool of CPU power. As long as a majority of CPU power is controlled by nodes that are not cooperating to attack the network, they'll generate the longest chain and outpace attackers. The network itself requires minimal structure. Messages are broadcast on a best effort basis, and nodes can leave and rejoin the network at will, accepting the longest proof-of-work chain as proof of what happened while they were gone.

CHAPTER 3

BITCOIN FIXES THIS

"I don't believe we shall ever have a good money again before we take the thing out of the hands of government, that is, we can't take it violently out of the hands of government, all we can do is by some sly roundabout way introduce something they can't stop."

F. A. Hayek, 1984

Taking Back Control

Prior to Bitcoin, digital payments could be (and still are) conducted online, but they require a trusted third party. This works fine if the trusted third party is trustworthy, but all experience has shown they are not. Payments are denied if the trusted third party, or a government, doesn't like what you are buying, or if you protest the wrong cause, or say the wrong thing, or don't take a vaccine, etc. etc. You could simply avoid shopping online, but that is severely restrictive and, these days, even most face-to-face payments are made electronically, thus requiring that dreaded third party. Even though relying on a third party to facilitate transactions is not desirable, there was simply no choice.

Until January 3, 2009.

That is when an anonymous computer coder gave Bitcoin to the world. Bitcoin takes digital buying and selling out of the control of a third party and puts it back in control of the buyer and seller. It takes the best of physical cash and puts it online, removing the dangers of cash payments. It provides the safety of distance but allows immediate settlement and avoids the risk of blocked payments, surveillance, and currency manipulation. Like cash (and gold), Bitcoin is a bearer instrument; he who holds it, owns it. And if that's not enough, the monetary policy of Bitcoin is as fair and just as can be conceived by men (Table 1).

Bitcoin	Dollar
Digital commodity	Fiat
Fixed supply	Infinite supply
Backed by energy	Backed by violence (i.e., law)
Native to the internet (digital)	Physical (digital requires trusted third party)
Audited every 10 minutes	Fed has never been audited
Fair distribution	Unfair distribution / Cantillon Effect
Cannot be debased	Debased by inflating money supply
Decentralized control	Centralized in Federal Reserve
Rules-based protocol	Controlled by central bank 'elites'
Deflationary forever	Inflationary by design
Becomes more valuable over time	Becomes less valuable over time
Salability scale: 10^8	Salability scale: 10^4

Table 1: Monetary Policies of Dollar vs Bitcoin

Bitcoin cannot be devalued by endless printing. Bitcoin is a voluntary system that is not forced upon anyone by coercion. The distribution of new bitcoins is just and fair and equitable. Bitcoins cannot be looted, nor can transactions be thwarted. Bitcoin is an open and transparent system that encourages cooperation and peace. It is the polar opposite of fiat money. So, can Bitcoin fix our money problems?

Digital Scarcity

Bitcoin is a digital money created in 2008 by an unknown creator going by the pseudonym Satoshi Nakamoto. Creating a workable, digital cash was something of a holy grail for computer programmers. The "problem" with computers is that all files and data are easily copied. When you send someone an email, for example, you actually send them a copy, and you also retain a copy on your device. If you send a photo to Facebook, the original stays on your device while a copy goes to Facebook. And anyone can download many copies after that. So how do you create a digital file that cannot be copied? Satoshi solved the problem using a combination of new technologies, including digital signatures, hashing, public/private key cryptography, decentralized networks, and a blockchain. The details are fairly complicated but suffice it to say it has worked for over 17 years, and the chain will soon have more than 1 million blocks. Satoshi's great accomplishment was creating *digital scarcity*. Bitcoin is the perfect money because it is perfectly scarce, durable, and digital.

Bitcoin is thus a digital commodity. Like gold, wheat, and other commodities, Bitcoin is controlled by no one and available to everyone. Unlike physical commodities, Bitcoin is programmatically *inelastic*, which means the supply is fixed (at 21 million bitcoins) and cannot be altered. If demand for gold or wheat increases, the production of gold or wheat will also increase to meet new demand. As demand goes up, prices go up, but as supply saturates demand, prices go down again. Since Bitcoin is inelastic, supply *cannot* be increased no matter how much the demand goes up. The only thing that can change is the price. This makes the Bitcoin price volatile, but only during its current 'discovery phase.' The volatility in Bitcoin's price will lessen as adoption increases, but its price in fiat will always go up because the value of fiat will always go down.

A Fair and Equitable Distribution
Everyone can participate in the Bitcoin ecosystem; all it takes is a phone or a computer. Just as anyone can mine gold or grow wheat, anyone can mine bitcoins. These days it takes a special computer, but in the early days any computer would do. Satoshi did not treat himself to a pile of bitcoins when it was launched, he mined them just like everyone else. Most of the other cryptocurrency copycats are launched with a pre-mine, a stash of coins given to the creators before the token goes public. These are scams designed to make the creators rich at the expense of gullible buyers. When Vitalik Buterin launched Ethereum, for example, he kept 60% of the total supply for himself. Bitcoin is the only true digital commodity money with a fair and equitable distribution. Also, the Bitcoin network has no knowledge of a person's gender, age, race, religion, or any other characteristic; Bitcoin literally cannot discriminate. Anyone can buy a computer and start mining bitcoins. They only need to buy the equipment and pay the electric bill. Anyone can download a wallet and start sending and receiving bitcoins. Anyone can run a node and enforce the rules of Bitcoin. None of these actions require permission from anyone. Traditional banking, on the other hand, routinely discriminates based on any number of factors, some within your control, others not.

By contrast, US dollars enter existence not through hard labor or expensive mining but through the stroke of a banker's digital pen. When the Federal Reserve expands the money supply, whether to avert financial collapse in 2008 or to fund unprecedented stimulus in 2020, the trillions of new dollars created do not disperse evenly across society like rain on a field. Instead, they cascade through specific channels: first to financiers and commercial banks, then to government bondholders, large corporations, and asset markets. By the time these dollars reach the paychecks of ordinary workers, the prices of homes, stocks, and commodities have already risen; everything on the shelf is already more expensive. By the Fed's own calculation, asset appreciation precedes CPI[1] inflation by 12-18 months. This sequential change in prices – first as asset appreciation, then as price inflation – was observed nearly three centuries ago by the economist Richard Cantillon. It remains one of modern finance's most enduring inequities. Between 2008 and 2022, the M2 money supply tripled, fueling asset bubbles that enriched the top 1 percent while median real wages stagnated (see graphs in Chapter 2). The bottom half of American households, holding just 2 percent of national wealth, bore the erosion of their savings as a hidden tax. A tax not levied by legislation, but by the very structure of fiat monetary issuance.

Bitcoin was conceived in the aftermath of the 2008 financial crisis and was launched in January 2009. The very first block mined, called the Genesis Block, enshrined a news article headline that became the financial shot heard 'round the world: *"The Times 03/Jan/2009 Chancellor on brink of second bailout for banks."* (Figure 1). With this pronouncement, the purpose of Bitcoin was made clear: to be a fair monetary system to counter the injustices of fiat banking.

Bitcoin operates under rules etched in code and enforced by a global network of computers. No central authority governs its supply and issuance; instead, these policies are written into its protocol and, being dispersed across thousands of computers globally, are nearly as unyielding to change as a law of physics. New bitcoins emerge not from privilege but from proof-of-work: miners compete to solve cryptographic puzzles, expending

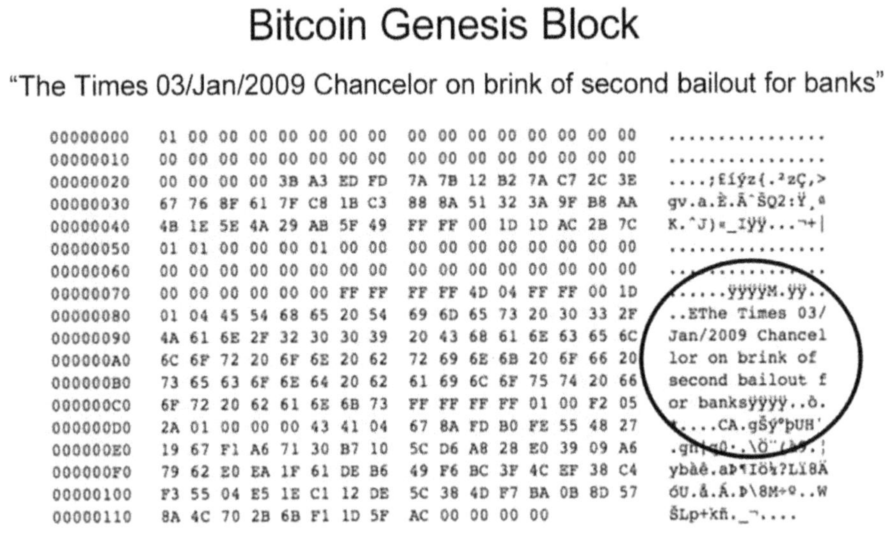

Bitcoin Genesis Block

"The Times 03/Jan/2009 Chancelor on brink of second bailout for banks"

Figure 1: Message in the Genesis Block

---

[1] CPI is consumer price index, a measure of inflation as reported by the central bank. Needless to say, the calculation methods are questionable.

electricity and computational power in a contest open to anyone with the desire to participate. Every ten minutes, some miner claims the block reward as compensation for their efforts. Yes, early participants gained far more bitcoins while expending far less energy or paying a far cheaper price, but they also bore considerably more uncertainty. The success of Bitcoin was uncertain, and they could have lost everything. Bitcoin is much more expensive for us joining the party later, but it's also far less risky. It is now clear that Bitcoin is here to stay, and its protocol guarantees no institution can accelerate issuance, and no policymaker can grant favors. The system demands real-world energy for every new coin, anchoring value in effort rather than proximity to power.

The contrast between the issuance of dollars and bitcoins is not merely technical but philosophical. Where fiat currency flows first to the connected – Wall Street trading desks, government contractors, and financiers – Bitcoin distributes new supply through a meritoc-racy of computation, blind to identity or influence. The Cantillon Effect thrives in opacity and discretion; Bitcoin neutralizes it with transparency and immutability. One system inflates assets before wages, widening inequality with each cycle of expansion. The other imposes scarcity by design, rewarding foresight and resilience while protecting long-term savers from dilution. Neither eliminates human disparity; wealth still concentrates among those who act early or allocate wisely, but Bitcoin removes the institutional thumb from the scale. For the first time since the age of gold, a form of money operates under rules that cannot be rewritten by committee. Rules enforced not by trust in men but by mathematics in cyberspace. Bitcoin has rules without rulers.

A Fixed Issuance & Supply

The Bitcoin protocol ensures that coins will be mined at a fixed pace, so coin production is entirely predictable. Not only is bitcoin issuance predictable and fair, but the issuance policy guarantees no more than 21 million bitcoins will ever be made. The details will be explained in the next chapter, but essentially the number of coins produced is cut in half every four years until eventually no more coins are produced at all. Production goes to zero at 21 million coins sometime in the year 2140. Because of this halving policy, bitcoin inflation was ini-tially very high but rapidly declined; almost 95% of the coins have already been mined. As of the last halving (April 20, 2024), the inflation rate of Bitcoin is about 1.7%. This is lower than the inflation rate of gold and lower than the Fed's target 2% inflation rate of dollars. (The actual inflation rate of dollars has averaged 6% since 1971). While bitcoin production will eventually go to zero, dollar production is headed to infinity (Figure 2).

Backed by Energy

The dollar was once backed by gold, and this kept the value of the dollar steady for many decades. Today, the dollar is backed by nothing.[2] Bitcoin is backed by energy.

---

[2] Actually, the dollar is backed by violence. If you fail to pay your taxes, men with guns will seize you. If a country attempts to trade outside dollars, we send troops to remind them of Bretton-Woods. Besides, war is good business for bankers, who throughout history have often funded both sides of a

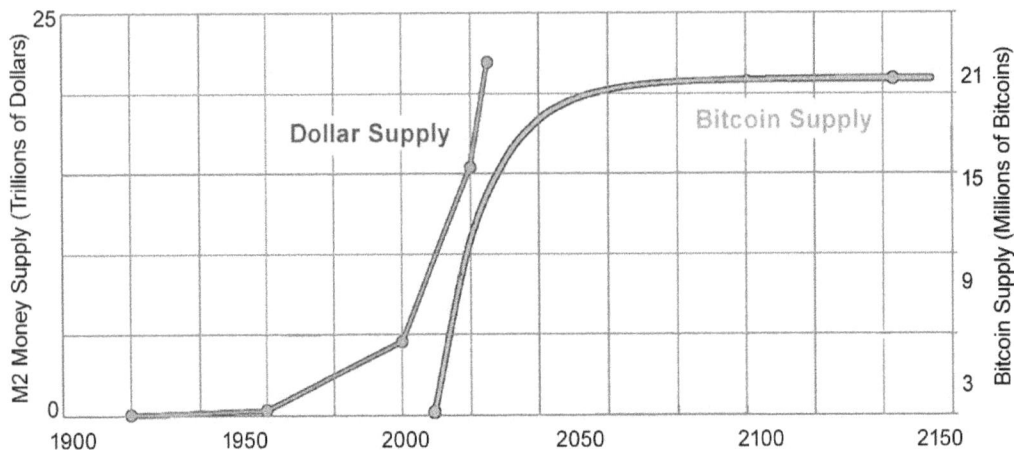

Figure 2: Supply of Dollars and Bitcoins

Henry Ford, creator of the assembly line and Ford Motor Company, once proposed an energy-backed form of money. Unfortunately, the technology didn't exist to make it possible. Today, it is possible and is achieved by Bitcoin, which uses electricity to mine new coins and to secure the blockchain (the public record of all Bitcoin transactions). So many computers are mining bitcoins today that the Bitcoin network is the most powerful computing network on Earth, by many orders of magnitude. To mine just one bitcoin a day requires nearly 36 megawatt-hours of electricity, enough to power 29,000 single-family homes. At the current issuance rate, 450 bitcoins are mined each day, consuming over 16,000 megawatt-hours per day. This sounds like a lot of electricity – and it is – but consider that 6 *million* megawatt-hours are consumed in the United States every Christmas season just for holiday lights. That's about 100,000 megawatt-hours per day for Christmas lights and 16,000 megawatt-hours per day for Bitcoin.

Even though we consume less electricity securing the Bitcoin network than we spend on Christmas lights, the amount of energy it would take to successfully attack the network is beyond the reach of even nation-states. It would take tens of thousands of megawatt-hours – basically what it takes to power New York City for a day – to reverse a single bitcoin transaction. A sustained attack would be required to cripple the network, and that is energetically impossible. Henry Ford would be so pleased.

The Federal Reserve was still new when Ford was alive, but he understood well the problems associated with a central bank. He once mused that *"it is well enough that people of the nation do not understand our banking and money system, for if they did, I believe there would be a revolution before tomorrow morning."*

---

conflict. The military-industrial complex has kept the United States in continual war for nearly a century, resulting in untold death, destruction, and global misery all to prop up the Almighty Dollar.

It has taken a long time, but the people of the nation are beginning to understand, and the revolution is now underway. Happily, it has thus far been a peaceful revolution.

Winner Take All

Money is a winner-take-all game, which is why most countries don't have two currencies simultaneously. When two monies do compete, history shows that the hardest[3] money always wins in a free market, since the hardest money performs the primary functions of money best, particularly being the best store of value and medium of exchange. If those two win, unit of account will follow.

In 2009, it was uncertain if Bitcoin would succeed. Other digital monies had been tried and failed. Yet Bitcoin was different. It was programmed to be the hardest money in the world and completely free of any central control. Its monetary policy was fair and just, and the code seemed promising. Soon after it launched in January 2009, Satoshi predicted that *"in 20 years there will either be very large transaction volume or no volume."*

Today, it is obvious there is "very large transaction volume" on Bitcoin. The network has a market cap of $2 trillion. Yet Bitcoin is still in its infancy. The same number of people

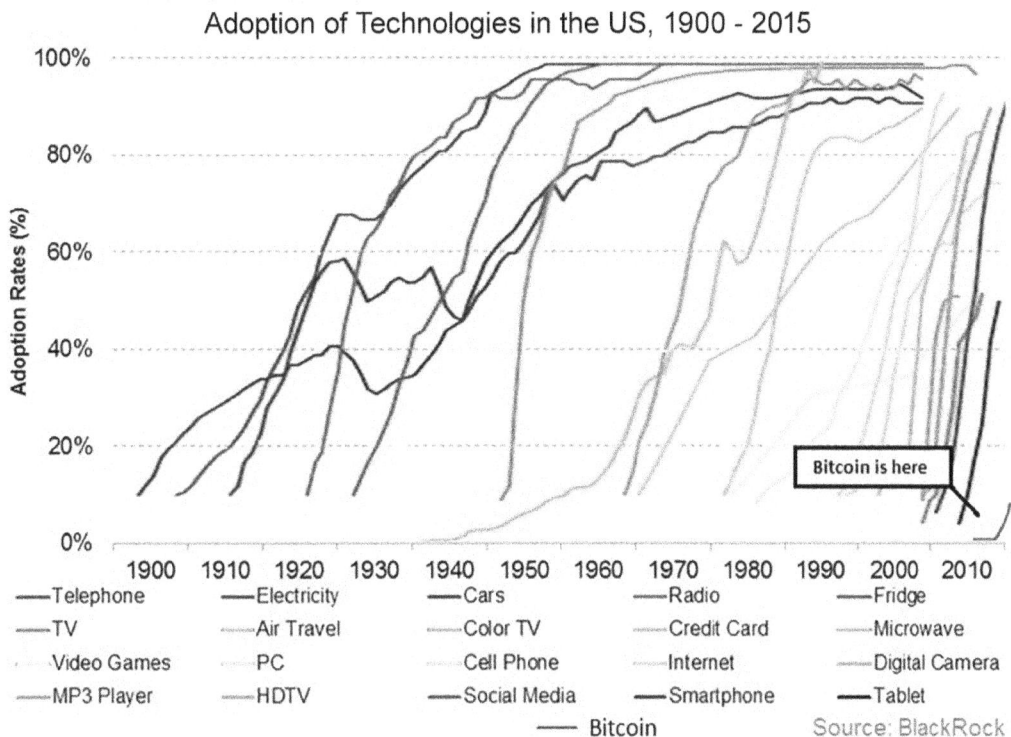

Figure 3: Adoption Curves for Various Technologies3

---

[3] A hard money is one that is hard to create. An easy money is one that is easy to create.

are using Bitcoin today as were using the internet in the late '90's, but Bitcoin adoption is growing *faster* than the internet did. Global adoption of Bitcoin has increased annually by 1000% for the past several years. In 2021, El Salvador made Bitcoin legal tender. In the United States, an estimated 50-65 million Americans own some bitcoin and in 2024, two presidential candidates and other prominent politicians spoke at the largest Bitcoin Conference in America. In January 2024, nine bitcoin ETF's were launched and were the most successful ETF launches in history. As for the computer code itself, almost 1 million blocks have been added to the blockchain with near flawless execution (there was one glitch in 2010 and another in 2013). The hashrate has risen steadily and continues to reach all-time highs. (The hashrate is a measure of network health). The network has never been hacked or forcibly stopped. It is safe to say that Bitcoin is succeeding while the dollar is collapsing and it is not a stretch to conclude that Bitcoin may one day replace the dollar as the global reserve currency or perhaps be used to back the dollar as gold once did. When Bitcoin replaces the dollar, it will usher in a new 'golden' era of peace and prosperity because Bitcoin incentivizes cooperation rather than conflict.

Money is a moral issue. The dollar is an immoral monetary system, and Bitcoin is an ethical alternative. Using Bitcoin whenever possible is a moral imperative.

As a technology, Bitcoin seems to be working. At this point, barring an unforeseen black swan event, it seems inevitable that Bitcoin will win because it is a superior form of money. Just as the car replaced the horse and the lightbulb replaced candles, newer and better technologies always replace older, inferior technologies. While cars impacted travel and lightbulbs impacted lighting, money impacts everything. We can only imagine the tranquility and riches that a world on a sound money standard can experience.

To me, Bitcoin is not simply about better tech or an investment or about making money. Money is a moral issue. I believe the dollar is an immoral monetary system, and Bitcoin is an ethical alternative. Using Bitcoin whenever possible is a moral imperative. This is why I am "all in" on Bitcoin. I do recognize – and you should, too – that Bitcoin is still a new technology and may yet fail, but I fight every day for it to succeed because right now it is our best hope against immoral money and financial tyranny. If Bitcoin goes to zero, I will sleep well knowing I fought the best fight available to me, for myself, my children, and my grandchildren.

CHAPTER 4

HOW BITCOIN WORKS

"I've developed a new open source P2P e-cash system called Bitcoin. It's completely decentralized, with no central server or trusted parties, because everything is based on crypto proof instead of trust."

Satoshi Nakamoto, 2008

No Bitcoins in Bitcoin

Although the pseudonymous creator of Bitcoin, Satoshi Nakamoto, used the word "Bitcoin" in his white paper (but only in the title), the bitcoin software doesn't actually recognize *bitcoins*. The native unit in the protocol is one hundred-millionth of what we call a bitcoin. All transactions, outputs, and balances are calculated and stored entirely in whole integers of this tiny unit. This unit was first called a 'satoshi' back in 2009 by Hal Finney to honor Bitcoin's creator. Hal Finney was the first person besides

Figure 1: Famous Tweet from Hal Finney

Nakamoto to run the bitcoin protocol and the first to conduct transactions with Nakamoto. Thus, a bitcoin is actually 100,000,000 satoshis and if you send 1 BTC to another person the protocol in fact sends 100,000,000 sats under the hood. And, yeah, *sats* is short for satoshis.

The three participants in the Bitcoin network are miners, nodes, and wallets. Let's explore these components and what they do. All run software, so let's start there.

Open-source vs Proprietary Code

Computer software can be either *open-source* or *closed-source* (proprietary). With open-source software, the code is freely available to the public; most are deposited to github.com. Thus, the software can be scrutinized by any interested computer programmer or hacker to look for bugs or secret backdoors. The code for proprietary software, used for example in popular voting machines, is not viewable to the public.

Bitcoin is an open-source project. The software is not only viewable, but any programmer can alter the code and/or suggest changes to the Bitcoin community. These suggested changes are called Bitcoin Improvement Proposals (BIPs). One of the most famous BIPs is BIP-039 which allows bitcoin private keys to be converted into a short list of English words (usually 12). This makes it much easier and safer to store private keys and back them up. Most BIPs only affect ancillary software, like wallets, or require only minor changes to Bitcoin. Any changes to the consensus rules (e.g., changing the maximum number of coins) would require a majority of node runners to update to the new version of the software, something they will not do if it compromises the integrity of the system and puts their wealth in jeopardy.

Public / Private Keys & Addresses

To access your bank account online, you need an account number and a password. However, there are no banks in Bitcoin; you access your money using wallets. A Bitcoin wallet uses a *public key* and *private key* instead of an account number and password. If you want someone to send money to your bank account, you give them your account number, but you would never share your password. Likewise, if you want to receive bitcoin, you may share your public key but never your private key. Anyone with access to your private key can take your bitcoin.

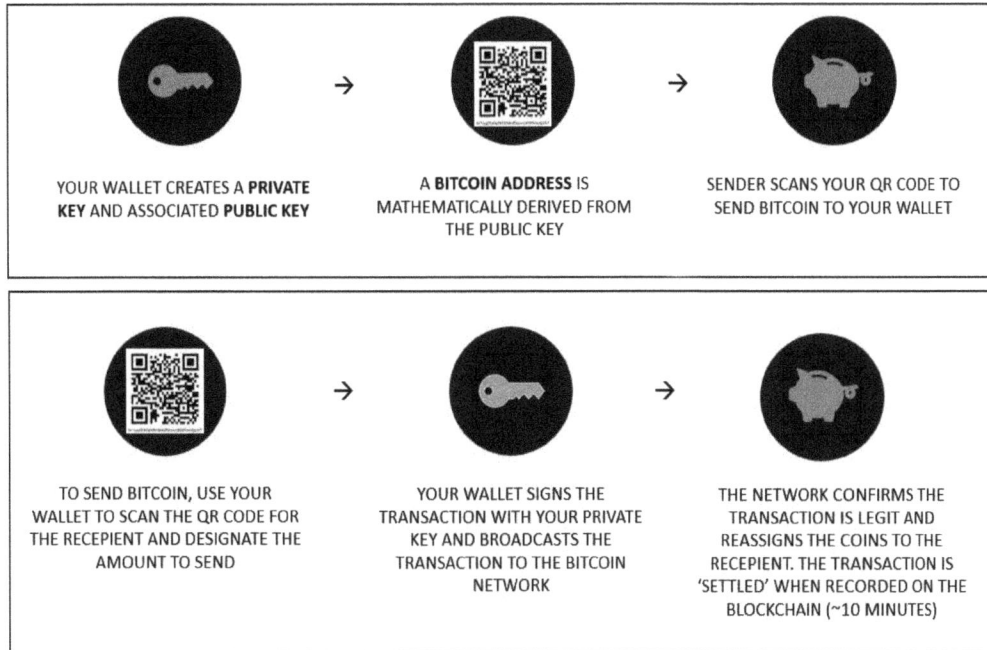

Figure 2: Sending & Receiving Bitcoin Transactions

Transactions are the transfer of sats from one wallet to another. To send someone sats, they must provide you with a receive address. Their receive addresses are derived from their public key. To initiate the transaction, you must sign it with your private key, proving you own the sats you are attempting to send. You also pay a small transaction fee to the miner to have them include your transaction in a block.

You can create your own keys by rolling dice or flipping coins, but usually we let the Bitcoin wallet create the keys for us. There are bitcoin wallets available for computers (Windows, Apple, and Linux) and for phones (Android and iOS). Once installed, these wallets will create a private key and a mathematically related public key for you.

Although you can share your public key, you usually won't. Instead, to receive sats you share a receive address derived from your public key. A new bitcoin address is generated for every transaction. Examples of an actual private key, public key, and bitcoin address are shown in Figure 3.

Bitcoin addresses are usually converted to QR codes by your wallet. You simply present the QR code to the sender and they can send you bitcoin. Or you can scan someone else's QR code to send them bitcoin. You can even send the QR code by text or email or post it on social media.

Nodes

Satoshi described Bitcoin as a "peer-to-peer electronic cash system." Each owner transfers money to the next owner by digitally signing a transaction with their private key.

Private key: E9873D79C6D87DC0FB6A577863338953213303DA61F20BD67FC233AA33263

Public key: FC7492E739D810291293098B38A74F0B82901369D937A798E4898D93987C9

Bitcoin address: bc1qyg0fck9g5640dwksff3scvvc50r65zeutpz0r7

Figure 3: A Bitcoin Private Key, Public Key, and Receive Address

Transactions are created and signed by Bitcoin *wallets*. The signed transaction is broadcast to computers called *nodes* that make up the peer-to-peer network. As of 2025, there are about 25,000 nodes visible on the internet scattered across the globe.

New transactions are broadcast to the network of nodes and stored temporarily in their memory. This waiting room for transactions is called the *mempool*. Computers called *miners* se-

bc1qyg0fck9g5640dwksff3scvvc50r65zeutpz0r7

lect transactions (usually based on a fee) from the mempool to put into a block and then perform work to get that block into the blockchain. Basically, miners do the hard work of building and securing the blockchain, and nodes keep the miners and wallets honest (because they will reject any blocks or transactions that do not follow the rules).

Nodes and miners are just computers. Anyone can run them. Because nodes and miners are scattered all over the planet and no single computer is in charge, the Bitcoin network is *decentralized*. That means there is no single point of failure and no central authority to be coerced. This is what makes Bitcoin a digital commodity.

Hash Functions

A *hash* is a mathematical function that takes any size input and produces an output of a fixed, predetermined size. The SHA256 function used by Bitcoin produces a 256-bit output. You can input a single character or an entire novel into the function, and it will produce a 256-bit output that is unique to each input. This is important: no two inputs will ever produce the same output. Equally important, the hash is a one-way function: you cannot determine the original input data from the hash output.

In the Bitcoin program, the 256-bit binary number is converted to a hexadecimal number, which will always be 64 characters long. The input 'Blue Ridge Bitcoin" results in the following 64-character hexadecimal string:

84918d83f85d7b8e201929857d2813476e145f1db6e6c7a93f3593e4c52a9aa2

Changing the input by adding "1" (i.e., Blue Ridge Bitcoin 1) results in:

3620456c2bfe10d5d11c536a753a1dee5084eac577f1852a75b2e39855ee5f6a

As you can see, altering the input even slightly completely changes the output. And again, there is no way to reverse the process to determine the input from the output.

Blocks

Blocks are collections of transactions to be added to the blockchain. Every block contains at least one transaction: the *coinbase* transaction which mints new bitcoins. Most blocks also contain hundreds or thousands of additional transactions, each one a tiny note saying something like *Alice sends Bob 50,000 sats*. Block size is limited to 1 MB, (expanded to about 4 MB via the SegWit upgrade), balancing security and scalability. Each transaction in the block is hashed to create a Merkle tree, and all are hashed together to create the Merkle root. Each block has a header, which contains the Merkle root of hashed transactions, a timestamp, the nonce for mining, and importantly, the hash of the previous block's header. This is what links the blocks together into a chain.

Blockchain

The blockchain is a distributed digital ledger that records all transactions in a chronological chain of immutable blocks. It is distributed because every full node running Bitcoin has its own copy. They are not copies in a traditional sense, but they are all identical due to *consensus*; since all nodes are following the same rules, they all come to the same conclusion regarding the state of the chain.

As noted above, each block contains a header with metadata including the hash of the previous block's header. Since every block contains a hash of the previous block's header, they are cryptographically linked to form a permanent chain. Attempting to alter any block would require recomputing all subsequent hashes, rendering it computationally infeasible. New blocks are appended to the chain approximately every 10 minutes. The blockchain's transparency allows anyone to verify the entire transaction history, providing a tamper-proof audit trail that underpins Bitcoin's trustless verification model.

Fun fact: Satoshi never used the word blockchain in the Bitcoin White Paper and some people prefer the term *timechain* because Satoshi emphasized that blocks were time-stamping transactions to create an unalterable history. The Bitcoin blockchain/timechain can be viewed and explored using the QR codes in Figure 4.

Mining

The word "miner" can refer to either the machine performing hashes are the humans that own and operate those machines. Miners (the computers) are extremely powerful machines that perform trillions of hashes per second. They are called *miners* because they create new bitcoins like mining for gold; however, a better term would be simply *hashers*. These computers consume large amounts of energy to build the blockchain, but the energy they consume is rather green. Miners (the people) are incentivized to find the cheapest energy possible and that is usually stranded energy or energy that would normally be wasted (e.g. methane flares). Because bitcoin miners (the machines and the people) are portable, they can go wherever cheap, abundant energy is located.

Figure 4: The Blockchain

timechaincalendar.com

mempool.space

The Difficulty

The *difficulty* is a target number. Miners must find a block hash below this target number to win a block. It's like pulling a number from a hat. Imagine there are 1 million numbered papers in a hat and you have to pull a number less than 100,000. You pull out a number and if it's greater than 100,000 you put it back in and try again. You keep trying until you pull a number less than 100,000, in which case you win. The difficulty is set to make sure you win once every 10 minutes, on average.

Now imagine 10 people join you in pulling numbers from the hat. Now that 10 times more effort is being put to the task, it is likely that a winning number will be found by someone in under 10 minutes on average, like maybe in 8 minutes on average. In that case, the difficulty will be adjusted to make winning harder. For example, the target might be lowered to 80,000. Now imagine 5 people leave the game; pulling a winning number could take longer, like 12 minutes on average. The target would be increased (to say 90,000) to make it take 10 minutes again. This process of making the challenge harder or easier depending on the number of people playing is called the *difficulty adjustment* and it results in a winner being declared at a regular interval (such as every 10 minutes, on average).

In Bitcoin, rather than people pulling numbers from a hat, you have computers hashing blocks. Recall that a hash produces a 64-character hexadecimal number and that even a slight change in the input produces a completely different output. The smaller the hexadecimal number, the more leading zeros it will contain. The difficulty target determines the number of leading zeros required to win a block. When mining, the Bitcoin program uses the block header as input for SHA256 hashing. If the hash is bigger than the target, a number is appended to the header, and it is re-hashed (the number appended is called a *nonce*, for 'number used once'). If the result is still bigger than the target, the nonce is changed and the header is hashed again. This process is repeated until a hash is obtained that is below the target value, in which case the miner wins the block.

If a lot of miners join the network, blocks will be found too quickly so the difficulty is increased. If many miners leave the network, as happened in 2021 when China banned Bitcoin mining, the difficulty is decreased so blocks are not added too slowly. The difficulty is programmatically adjusted every 2,016 blocks (approximately every two weeks). The difficulty adjustment in the Bitcoin protocol guarantees that coins will be minted at a fixed pace regardless of the number of miners on the network. In this way, Bitcoin issuance is completely predictable.

As stated above, the target number is reflected in the number of leading zeros the hexadecimal hash output contains. At the time of writing, the current difficulty is 101 trillion, meaning it is about 101 trillion times harder to find a block today than when Satoshi mined the first block (when the difficulty was 1). The current target requires a hash with 19 leading zeros to win a block. The hash for block 869,206 is

Block 869206: **0000000000000000000**102415becd6036ecf68ef0e516715c222bbbb7bb3d182

For comparison, the hash for block 5 only has 8 zeros, but Satoshi was the only one mining bitcoins at that time.

Block 5: **00000000**9b7262315dbf071787ad3656097b892abffd1f95a1a022f896f533fc

How hard is it to find a winning hash today? The entire bitcoin network (i.e., all miners combined) is currently performing more than 1 ZH/s in order for someone to win a block in 10 minutes. One zeta hash is 1,000,000,000,000,000,000,000 hashes. This is how many hashes must be done *every second* to find a winning hash within 600 seconds (i.e., 10 minutes).

There are many miners around the world competing for the next block. The first miner to produce a hash with the requisite number of leading zeros wins the block. Their block is added to the blockchain, and they receive a payout called the *block reward*. The block reward consists of fees paid by users to include their transactions in the block and the *block subsidy* of newly minted bitcoins. The subsidy, currently 3.125 BTC, is reduced by half every 210,000 blocks (roughly every 4 years) during an event called the *halving*. Each 4-year period is called an *epoch*. We are presently in Bitcoin's fifth epoch until about April 2028. Miners are strongly incentivized to follow the rules because having a single block rejected currently costs around $300,000 in lost revenue.

Recall the hash from "Blue Ridge Bitcoin" was

84918d83f85d7b8e201929857d2813476e145f1db6e6c7a93f3593e4c52a9aa2 and adding the nonce "1" (i.e., Blue Ridge Bitcoin 1) changes it to

3620456c2bfe10d5d11c536a753a1dee5084eac577f1852a75b2e39855ee5f6a

The nonce "10" creates a hash with one leading zero:

0dc79b2fc9075e36dfa71bbc770d80c4970ef72ff11a0500ecb4b59cec672dc6

The nonce "22" creates a hash with two leading zeros:

00ef3c1bbe4176a7298968c28846f9dadd8edfc3d3e914ad62017f31c762661a

How many hashes must be performed to find an output with 19 leading zeros? Statistically, it should take about 600 billion billion hashes!

Hashing is used in Bitcoin mining because it is somewhat like a Sudoku puzzle; it can be extremely difficult to solve the puzzle but easy to confirm a winning solution. Likewise, it is difficult for a miner to find a winning hash, but it's easy for distributed nodes to confirm a winning hash.

Proof of Work

Bitcoin employs a proof-of-work (PoW) consensus algorithm to secure the network and to achieve agreement on the blockchain's state among distributed nodes. Finding a hash of requisite difficulty is called *proof of work* because it can only be accomplished by expending real-world energy, i.e., electricity. As discussed in the previous chapter, a miner would have to consume about 36 megawatt-hours of electricity to find just one block per day. In other words, if you find one block per day, it proves you expended about 36 megawatt-hours to do so; it is proof you did the work.

Sudoku puzzle

This proof-of-work mechanism, i.e., having to consume energy to find a valid block, is what links the virtual currency to the real (physical) world. If a bad actor tried to alter a past transaction, they would have to redo the PoW which would incur real-world energy costs. This mechanism incentivizes honest participation because trying to undo more than one block would be prohibitively expensive, even for nation-states, as an attacker would need to out-compute the majority of the network's hashrate for a sustained period of time. Consequently, the blockchain is an unalterable history of bitcoin transactions.

Bitcoin Wallets

In the physical world, a wallet is typically made of leather and stitching. It's a place to hold money as paper and plastic. However, Bitcoin wallets do not hold your sats; they store and manage the cryptographic keys that control your sats. The basic functions of a bitcoin wallet include

1. Generate and/or store your private and public keys.
2. Use your public key to generate receive addresses for receiving bitcoin.
3. Create transactions.
4. Use your private key to sign transactions for sending bitcoin.
5. Broadcast your transaction to the Bitcoin network.

Wallets have become increasingly sophisticated and easy to use over the years, but all are either *hot wallets* or *cold wallets*. A hot wallet is connected to the internet. There are hot wallets designed for laptops and computers, but most hot wallets are designed for cell phones. These wallets are typically used for small purchases, not long-term storage. Nearly all hot wallets these days are also lightning wallets, meaning they interact with the lightning network – a second layer – to perform extremely fast transactions (see Scaling Bitcoin below).

A cold wallet is one that never connects to the internet. These are used for long-term storage and storing large amounts of bitcoin. Think of a hot wallet as a checking account and a cold wallet as a savings account. Wallets can also be *custodial* or *self-custodial*. You hold your own Bitcoin keys with a self-custodial wallet. A custodial wallet is one where someone else holds your keys for you. This can be useful if you don't trust yourself to keep your bitcoins safe, but it comes with a third-party risk, which negates the point of using Bitcoin. As Bitcoiners say: "*Not your keys, not your coins.*" If you keep your bitcoins on the exchange where you bought them, you do not hold the keys; they do.

Seed Words

It is possible to create Bitcoin keys yourself, but usually you let the wallet do it for you. It begins with a large random number called the *entropy*. The entropy may be created by a random number generator in your device, but good wallets incorporate noise from the real world, such as thermal noise across a diode, or values from pixels in a photo. Most wallets use an entropy that is 128-bits long. Your private and public keys will be derived from this number, so you absolutely need a backup stored off the device.

1	umbrella
2	random
3	purpose
4	lift
5	rugged
6	season
7	mystery
8	airplane
9	jungle
10	silence
11	tool
12	energy

Attempting to write down this enormous number by hand practically guarantees a mistake, and yet you never want an electronic backup (not even a digital photo) because that can be hacked. To make backing up the entropy by hand easier, a Bitcoin Improvement Proposal (BIP) has been adopted by all modern wallets. BIP-39 takes this long, random number and transforms it into 12 English words, called *seed words* or a *seed phrase*. A seed phrase is much easier for a human to write down or even memorize. To back up your wallet, write down the seed words *by hand* and *in order*, and save them in a very safe place, like where you might store gold coins or expensive jewelry. If someone asks you for your seed words, say "NO" and *NEVER* give it to them, unless you want them to control your bitcoin. If you lose the words, you lose access to your money. If you lose your wallet (i.e., your phone or laptop), you can restore access to your money by importing the seed words into any wallet on another device.

An optional "13th word" called a *passphrase* can be added for an extra layer of protection. The passphrase can be a word or a sentence, and it can include letters, numbers, and special characters.

To reiterate, it is absolutely essential that you back up your wallet's private key (and passphrase if using one).[1] If you lose your phone or it's stolen, or if the computer holding your wallet crashes, you will *lose your bitcoin* if you do not have your private key backed up. With your private key, however, you can restore your wallet on a new device and regain access to your bitcoin.

Public keys are mathematically derived from private keys. Public keys can be shared, but doing so allows the other party to see your bitcoin balance and every transaction associated with your wallet. This could be desirable when full transparency is the goal (such as for churches or governments), but it obviously leaks privacy. To protect your privacy, you supply the sender with a receive address instead of the public key. Receive addresses are mathematically derived from public keys, and a new receive address will be generated for every transaction. So, receive addresses are mathematically derived from public keys, which are mathematically derived from private keys, which are mathematically derived from a random number, which is backed up as 12 English words called your seed words. All of this (except the backup) is managed by your wallet.

A Bitcoin wallet is the primary human interface with the Bitcoin network. Indeed, by simply memorizing 12 words you can hold enormous wealth in your head (called a brain wallet[2]). Try boarding a plane with $10 million in cash stuffed in a suitcase or crossing any port-of-entry with 5 pounds of gold.

One of my favorite movies is *The Sound of Music*. As the movie ends and the credits begin to roll, the von Trapp family is seen trekking over the snowy Alps to escape the Nazis. Each member of the family carries their essentials in a suitcase, but the family was forced to leave cars, bank accounts, and a mansion behind. For those escaping persecution today, they can carry *all* of their wealth in their heads, and this is yet another reason by Bitcoin is called freedom money.

Scaling Bitcoin Use & Adoption

The blockchain and all transactions within it are called the *base layer* of Bitcoin. Because blocks can only hold about 3,000 transactions and new blocks are added every 10 minutes, the base layer can only process about 5 transactions per second. The VISA network, by contrast, regularly processes 50 thousand transactions per second. Given the (intentionally) slow speed of the blockchain, how can Bitcoin scale to be a global monetary network?

---

[1] Bitcoiners often refer to the seed phrase as a *backup of your private key* but technically you are backing up the entropy from which the private key is derived.
[2] Brain wallets are recommended only for emergencies. Relying on a brain wallet means all your money can be lost by a brain injury, a stroke, or simply forgetting the words.

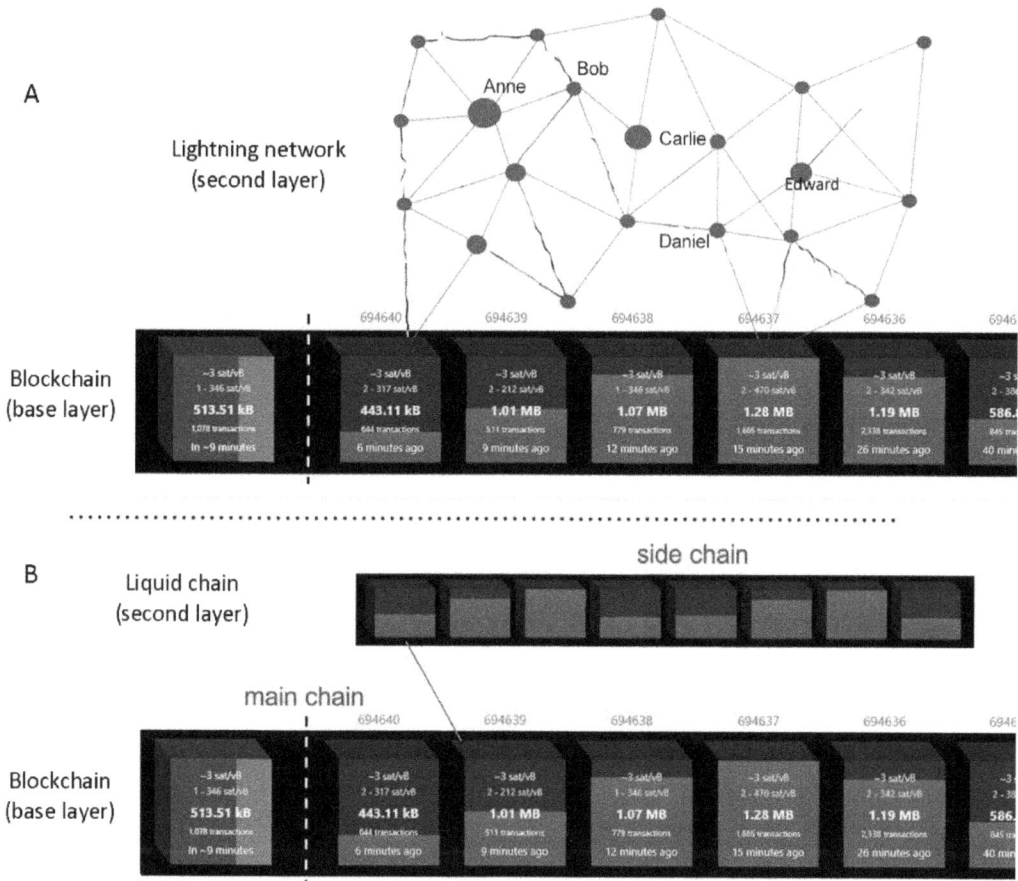

Figure 5: Second Layers

Bitcoin's base layer need not process every coffee purchase. *Second layer* solutions such as the *lightning network* and *side chains* have been developed for these speedier transactions (Figure 5). Second layers are pegged to the base layer to remain faithful to the monetary policies of Bitcoin. The lightning network enables speedy payments via channels connecting users. These days, lightning wallets open and manage these channels under the hood. The lightning network now routes over 5,000 BTC in liquidity across tens of thousands of channels, enabling instant, near-free micropayments while inheriting the base layer's final settlement guarantees. Millions of merchants already accept satoshis as readily as dollars, proving that a second-layer economy can thrive.

Side chains are separate blockchains that follow different rules but are pegged to the Bitcoin main chain. The Liquid side chain produces one block per minute and thus runs 10 times faster than the main chain. One L-BTC equals one BTC and lightning can run on top of the Liquid chain, too.

39

Beyond technology, scaling demands education and infrastructure that speak in local languages and to local needs. In the United States, where most people have bank accounts and fiat payments are easy, Bitcoin education must tackle rising inflation and deteriorating property rights. Convincing the American that Bitcoin is a better savings technology is key, and the importance of taking self-custody to avoid a future 6102-style attack. In nations with crumbling currencies, adoption is less about earning a yield and more about survival: protecting the Venezuelan mother from a 20% fee when receiving remittances, for instance. Stablecoins can bridge fiat on-ramps, but the endgame is always the same: an exit to sound money and self-custody. The global Bitcoin network grows by individuals choosing – one private key at a time – to opt out of systems that print tomorrow's lunch with today's keystrokes and step between people simply wanting to send value to each other, extracting a fee in the process.

The coming shift will be cultural, not just economic. Bitcoin is not just a new payment technology; it is a realignment of trust. Older generations lived in a high-trust society, but the abuses of fiat money (and a concurrent spiritual decline) have eroded that trust. Where central banks require faith in flawed humans, Bitcoin demands verification in code. Turns out, computer code is more trustworthy than people.

Global Bitcoin adoption (termed *hyperbitcoinization* by Max Keiser) will not arrive with a single nation-state decree, or Strategic Bitcoin Reserves; it will emerge when a critical mass of people, from New York to Nigeria, recognize that a 12-word seed phrase is freedom from financial slavery. I am bullish on the future because blocks will keep coming every ten minutes, indifferent to borders or bailouts, until the ledger beneath our feet becomes the heartbeat of a more fair and just economic order.

Tick tock, next block.

CHAPTER 5

GETTING STARTED WITH BITCOIN

"Getting off zero is the most important financial decision most people will ever make."

- Vijay Boyapati

A Deep Rabbit Hole

Discovering Bitcoin is described as falling down a rabbit hole. This metaphor, taken from *Alice in Wonderland*, aptly describes the journey. Bitcoin unveils a world you never saw before. A reality that once seen can never be unseen. To deeply examine Bitcoin is to trace a thread through the interwoven tapestry of history, economics, math, computer science, and even philosophy and religion.

Bitcoin is not merely a technical project; it is a human project. Your wallet, something that appears to be just another app on your phone, is working with nodes and miners around the globe to remove the centralized gatekeepers to your money. It replaces overlords, regulators, and bureaucrats with independence and personal responsibility. These tools offer something even more valuable than bitcoins; they offer dignity and sovereignty embedded in code.

You could read a hundred books, but the best way to discover Bitcoin is to own some and hold it in self-custody. The first time you send (or receive) sats free of third-party interference is simply magical. If your curiosity has brought you this far, you need to take the plunge and see how deep the rabbit hole really goes.

Am I Too Late?

Many people ask if they are too late to Bitcoin. While it is true that early adopters reaped significant gains, to ask the question reveals you still haven't completed the paradigm shift that is Bitcoin. When you buy bitcoin, you are not actually *buying* bitcoins; you are exchanging an inferior form of money for a superior form of money. One thousand sats may be worth 1 US dollar or 31 Turkish liras or 217 Venezuelan bolivars; your wealth is measured in how many sats you own, not their worth in fiat paper. There is a limited number of sats but an infinite supply of fiat.

You are not too late to Bitcoin. Those who exchanged dollars for sats at $100/coin, $1,000/coin, and $100,000/coin, will still be doing it at $1 million/coin and $10 million/coin. It's never too late to swap bad money for good money. Because Bitcoin is the hardest money to ever exist, most sound money advocates believe it's only a matter of time before Bitcoin dominates the world. Your children won't have to bother with an exchange; they will get paid in sats.

Getting & Keeping Bitcoin Safely

There are basically four ways to acquire bitcoin: 1) buy it from someone in person, 2) buy it from an exchange, 3) mine sats, or 4) earn sats. Buying from a stranger in person can be difficult to arrange and, frankly, pretty dangerous. Mining bitcoin these days takes expensive equipment to be competitive. I encourage everyone to mine bitcoin at home, but home mining is primarily to contribute decentralized hashrate to the network with no expectation of actually winning a block. Earning bitcoin is difficult because Bitcoin is not yet a universal medium of exchange. The number of merchants accepting bitcoin as payment is growing rapidly, so earning bitcoin by selling goods and services or even getting paid in bitcoin may be a viable option soon. But the most common way to acquire bitcoin for now is through an exchange.

When buying bitcoin – or rather, when exchanging dollars for bitcoin – it's crucial to choose a reputable exchange. Mt. Gox and FTX are two dramatic examples of exchanges that mishandled customers' funds, and many customers lost money. Coinbase is the most popular exchange, but not one that I would recommend at all. Personally, I consider Coinbase a Mt. Gox or FTX that just hasn't exploded yet.

Next Steps

Buying bitcoin is only the first step. The next step is to move your sats into a self-custodial wallet. Remember: *not your keys, not your coins*. Once you acquire a sizeable stack or simply have sats you don't intend to move for a long time, you should put those sats into cold storage. This entails moving your sats from a hot wallet to an air-gapped cold wallet.

After securing your bitcoin stack in self-custodial cold storage, you should consider running a Bitcoin node. Everyone with skin in the game should run a node. Not only can it increase your privacy, it also strengthens the network. Bitcoin is immune to policy changes (e.g., changing the 21 million cap) and hostile actors (including tyrannical governments) solely because it is *decentralized*, meaning thousands of nodes across the globe are running the code and connecting peer-to-peer. You can join that decentralized network and become a legitimate Guardian of the Timechain with only a small computer and a reliable internet connection.

Finally, to complete your Bitcoin journey, you should run a home miner. There are several home miners on the market that are easy to use. These little miners are not likely to win a block (though several have), but you will learn a lot about how Bitcoin actually works by running a node and a miner. It's fun just knowing your little miner is building block templates and hashing them along with the big boys.

On a more serious note, Bitcoin mining has become dangerously centralized in recent years because of those big boys. Every small miner adds hash power to the network and improves mining decentralization. Innovative entrepreneurs are even trying to incorporate home miners into appliances that need heat, such as water heaters and clothes dryers. Imagine if your water heater made you money instead of costing you money! Put on your sunglasses, the future is bright orange.

If you would like to accelerate the arrival of this future and need help buying bitcoin, using a wallet, securing your stack, or setting up a node, please reach out to Blue Ridge Bitcoin or another Bitcoin coaching service for help.

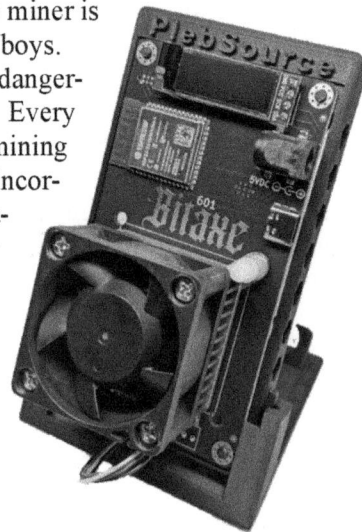

Resources for going deeper down the rabbit hole:
Recommended Books
The Bitcoin Standard, by Saifedean Ammous
Thank God for Bitcoin, by Breedlove et al.
Broken Money, by Lynn Alden
The Creature from Jekyll Island, by G. Edward Griffen
End the Fed, by Ron Paul
The Big Print, by Lawrence Lepard

Recommended YouTube channels
Simply Bitcoin
Prof St Onge
Swan Bitcoin
Bitcoin University
BTC Sessions
The Bitcoin Way

The Bitcoin White Paper

Published October 31, 2008

Bitcoin White Paper (bitcoin.org)

Bitcoin: A Peer-to-Peer Electronic Cash System

Satoshi Nakamoto
satoshin@gmx.com
www.bitcoin.org

Abstract. A purely peer-to-peer version of electronic cash would allow online payments to be sent directly from one party to another without going through a financial institution. Digital signatures provide part of the solution, but the main benefits are lost if a trusted third party is still required to prevent double-spending. We propose a solution to the double-spending problem using a peer-to-peer network. The network timestamps transactions by hashing them into an ongoing chain of hash-based proof-of-work, forming a record that cannot be changed without redoing the proof-of-work. The longest chain not only serves as proof of the sequence of events witnessed, but proof that it came from the largest pool of CPU power. As long as a majority of CPU power is controlled by nodes that are not cooperating to attack the network, they'll generate the longest chain and outpace attackers. The network itself requires minimal structure. Messages are broadcast on a best effort basis, and nodes can leave and rejoin the network at will, accepting the longest proof-of-work chain as proof of what happened while they were gone.

1. Introduction

Commerce on the Internet has come to rely almost exclusively on financial institutions serving as trusted third parties to process electronic payments. While the system works well enough for most transactions, it still suffers from the inherent weaknesses of the trust based model. Completely non-reversible transactions are not really possible, since financial institutions cannot avoid mediating disputes. The cost of mediation increases transaction costs, limiting the minimum practical transaction size and cutting off the possibility for small casual transactions, and there is a broader cost in the loss of ability to make non-reversible payments for non-reversible services. With the possibility of reversal, the need for trust spreads. Merchants must be wary of their customers, hassling them for more information than they would otherwise need. A certain percentage of fraud is accepted as unavoidable. These costs and payment uncertainties can be avoided in person by using physical currency, but no mechanism exists to make payments over a communications channel without a trusted party.

What is needed is an electronic payment system based on cryptographic proof instead of trust, allowing any two willing parties to transact directly with each other without the need for a trusted third party. Transactions that are computationally impractical to reverse would protect sellers from fraud, and routine escrow mechanisms could easily be implemented to protect buyers. In this paper, we propose a solution to the double-spending problem using a peer-to-peer distributed timestamp server to generate computational proof of the chronological order of transactions. The system is secure as long as honest nodes collectively control more CPU power than any cooperating group of attacker nodes.

1

2.   Transactions

We define an electronic coin as a chain of digital signatures. Each owner transfers the coin to the next by digitally signing a hash of the previous transaction and the public key of the next owner and adding these to the end of the coin. A payee can verify the signatures to verify the chain of ownership.

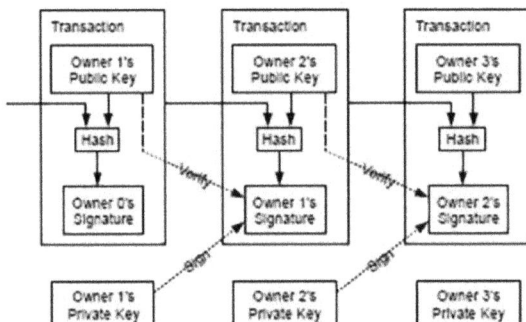

The problem of course is the payee can't verify that one of the owners did not double-spend the coin. A common solution is to introduce a trusted central authority, or mint, that checks every transaction for double spending. After each transaction, the coin must be returned to the mint to issue a new coin, and only coins issued directly from the mint are trusted not to be double-spent. The problem with this solution is that the fate of the entire money system depends on the company running the mint, with every transaction having to go through them, just like a bank.

We need a way for the payee to know that the previous owners did not sign any earlier transactions. For our purposes, the earliest transaction is the one that counts, so we don't care about later attempts to double-spend. The only way to confirm the absence of a transaction is to be aware of all transactions. In the mint based model, the mint was aware of all transactions and decided which arrived first. To accomplish this without a trusted party, transactions must be publicly announced [1], and we need a system for participants to agree on a single history of the order in which they were received. The payee needs proof that at the time of each transaction, the majority of nodes agreed it was the first received.

3.   Timestamp Server

The solution we propose begins with a timestamp server. A timestamp server works by taking a hash of a block of items to be timestamped and widely publishing the hash, such as in a newspaper or Usenet post [2-5]. The timestamp proves that the data must have existed at the time, obviously, in order to get into the hash. Each timestamp includes the previous timestamp in its hash, forming a chain, with each additional timestamp reinforcing the ones before it.

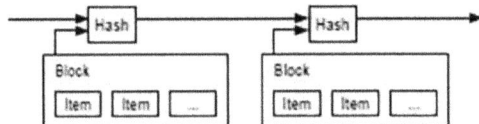

2

48

4. Proof-of-Work

To implement a distributed timestamp server on a peer-to-peer basis, we will need to use a proof-of-work system similar to Adam Back's Hashcash [6], rather than newspaper or Usenet posts. The proof-of-work involves scanning for a value that when hashed, such as with SHA-256, the hash begins with a number of zero bits. The average work required is exponential in the number of zero bits required and can be verified by executing a single hash.

For our timestamp network, we implement the proof-of-work by incrementing a nonce in the block until a value is found that gives the block's hash the required zero bits. Once the CPU effort has been expended to make it satisfy the proof-of-work, the block cannot be changed without redoing the work. As later blocks are chained after it, the work to change the block would include redoing all the blocks after it.

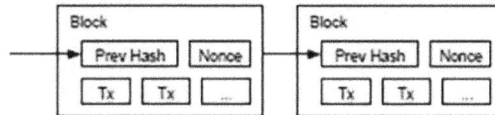

The proof-of-work also solves the problem of determining representation in majority decision making. If the majority were based on one-IP-address-one-vote, it could be subverted by anyone able to allocate many IPs. Proof-of-work is essentially one-CPU-one-vote. The majority decision is represented by the longest chain, which has the greatest proof-of-work effort invested in it. If a majority of CPU power is controlled by honest nodes, the honest chain will grow the fastest and outpace any competing chains. To modify a past block, an attacker would have to redo the proof-of-work of the block and all blocks after it and then catch up with and surpass the work of the honest nodes. We will show later that the probability of a slower attacker catching up diminishes exponentially as subsequent blocks are added.

To compensate for increasing hardware speed and varying interest in running nodes over time, the proof-of-work difficulty is determined by a moving average targeting an average number of blocks per hour. If they're generated too fast, the difficulty increases.

5. Network

The steps to run the network are as follows:

1) New transactions are broadcast to all nodes.
2) Each node collects new transactions into a block.
3) Each node works on finding a difficult proof-of-work for its block.
4) When a node finds a proof-of-work, it broadcasts the block to all nodes.
5) Nodes accept the block only if all transactions in it are valid and not already spent.
6) Nodes express their acceptance of the block by working on creating the next block in the chain, using the hash of the accepted block as the previous hash.

Nodes always consider the longest chain to be the correct one and will keep working on extending it. If two nodes broadcast different versions of the next block simultaneously, some nodes may receive one or the other first. In that case, they work on the first one they received, but save the other branch in case it becomes longer. The tie will be broken when the next proof-of-work is found and one branch becomes longer; the nodes that were working on the other branch will then switch to the longer one.

3

New transaction broadcasts do not necessarily need to reach all nodes. As long as they reach many nodes, they will get into a block before long. Block broadcasts are also tolerant of dropped messages. If a node does not receive a block, it will request it when it receives the next block and realizes it missed one.

6. Incentive

By convention, the first transaction in a block is a special transaction that starts a new coin owned by the creator of the block. This adds an incentive for nodes to support the network, and provides a way to initially distribute coins into circulation, since there is no central authority to issue them. The steady addition of a constant of amount of new coins is analogous to gold miners expending resources to add gold to circulation. In our case, it is CPU time and electricity that is expended.

The incentive can also be funded with transaction fees. If the output value of a transaction is less than its input value, the difference is a transaction fee that is added to the incentive value of the block containing the transaction. Once a predetermined number of coins have entered circulation, the incentive can transition entirely to transaction fees and be completely inflation free.

The incentive may help encourage nodes to stay honest. If a greedy attacker is able to assemble more CPU power than all the honest nodes, he would have to choose between using it to defraud people by stealing back his payments, or using it to generate new coins. He ought to find it more profitable to play by the rules, such rules that favour him with more new coins than everyone else combined, than to undermine the system and the validity of his own wealth.

7. Reclaiming Disk Space

Once the latest transaction in a coin is buried under enough blocks, the spent transactions before it can be discarded to save disk space. To facilitate this without breaking the block's hash, transactions are hashed in a Merkle Tree [7][2][5], with only the root included in the block's hash. Old blocks can then be compacted by stubbing off branches of the tree. The interior hashes do not need to be stored.

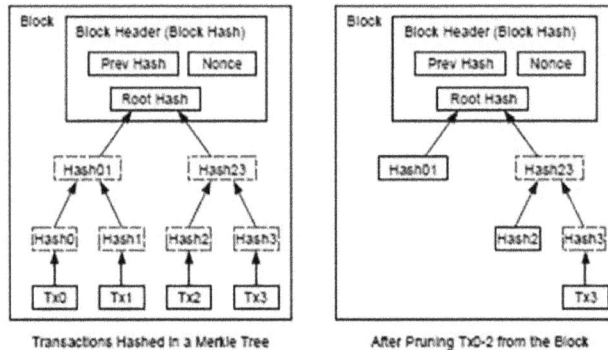

Transactions Hashed in a Merkle Tree   After Pruning Tx0-2 from the Block

A block header with no transactions would be about 80 bytes. If we suppose blocks are generated every 10 minutes, 80 bytes * 6 * 24 * 365 = 4.2MB per year. With computer systems typically selling with 2GB of RAM as of 2008, and Moore's Law predicting current growth of 1.2GB per year, storage should not be a problem even if the block headers must be kept in memory.

4

8.   Simplified Payment Verification

It is possible to verify payments without running a full network node. A user only needs to keep a copy of the block headers of the longest proof-of-work chain, which he can get by querying network nodes until he's convinced he has the longest chain, and obtain the Merkle branch linking the transaction to the block it's timestamped in. He can't check the transaction for himself, but by linking it to a place in the chain, he can see that a network node has accepted it, and blocks added after it further confirm the network has accepted it.

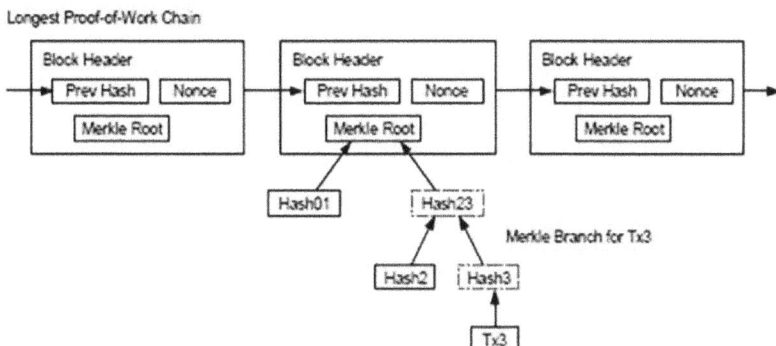

As such, the verification is reliable as long as honest nodes control the network, but is more vulnerable if the network is overpowered by an attacker. While network nodes can verify transactions for themselves, the simplified method can be fooled by an attacker's fabricated transactions for as long as the attacker can continue to overpower the network. One strategy to protect against this would be to accept alerts from network nodes when they detect an invalid block, prompting the user's software to download the full block and alerted transactions to confirm the inconsistency. Businesses that receive frequent payments will probably still want to run their own nodes for more independent security and quicker verification.

9.   Combining and Splitting Value

Although it would be possible to handle coins individually, it would be unwieldy to make a separate transaction for every cent in a transfer. To allow value to be split and combined, transactions contain multiple inputs and outputs. Normally there will be either a single input from a larger previous transaction or multiple inputs combining smaller amounts, and at most two outputs: one for the payment, and one returning the change, if any, back to the sender.

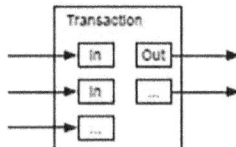

It should be noted that fan-out, where a transaction depends on several transactions, and those transactions depend on many more, is not a problem here. There is never the need to extract a complete standalone copy of a transaction's history.

5

10. Privacy

The traditional banking model achieves a level of privacy by limiting access to information to the parties involved and the trusted third party. The necessity to announce all transactions publicly precludes this method, but privacy can still be maintained by breaking the flow of information in another place: by keeping public keys anonymous. The public can see that someone is sending an amount to someone else, but without information linking the transaction to anyone. This is similar to the level of information released by stock exchanges, where the time and size of individual trades, the "tape", is made public, but without telling who the parties were.

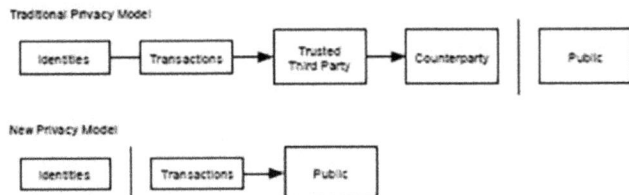

As an additional firewall, a new key pair should be used for each transaction to keep them from being linked to a common owner. Some linking is still unavoidable with multi-input transactions, which necessarily reveal that their inputs were owned by the same owner. The risk is that if the owner of a key is revealed, linking could reveal other transactions that belonged to the same owner.

11. Calculations

We consider the scenario of an attacker trying to generate an alternate chain faster than the honest chain. Even if this is accomplished, it does not throw the system open to arbitrary changes, such as creating value out of thin air or taking money that never belonged to the attacker. Nodes are not going to accept an invalid transaction as payment, and honest nodes will never accept a block containing them. An attacker can only try to change one of his own transactions to take back money he recently spent.

The race between the honest chain and an attacker chain can be characterized as a Binomial Random Walk. The success event is the honest chain being extended by one block, increasing its lead by +1, and the failure event is the attacker's chain being extended by one block, reducing the gap by -1.

The probability of an attacker catching up from a given deficit is analogous to a Gambler's Ruin problem. Suppose a gambler with unlimited credit starts at a deficit and plays potentially an infinite number of trials to try to reach breakeven. We can calculate the probability he ever reaches breakeven, or that an attacker ever catches up with the honest chain, as follows [8]:

p = probability an honest node finds the next block
q = probability the attacker finds the next block
q_z = probability the attacker will ever catch up from z blocks behind

$$q_z = \begin{cases} 1 & if\ p \le q \\ (q/p)^z & if\ p > q \end{cases}$$

6

Given our assumption that $p > q$, the probability drops exponentially as the number of blocks the attacker has to catch up with increases. With the odds against him, if he doesn't make a lucky lunge forward early on, his chances become vanishingly small as he falls further behind.

We now consider how long the recipient of a new transaction needs to wait before being sufficiently certain the sender can't change the transaction. We assume the sender is an attacker who wants to make the recipient believe he paid him for a while, then switch it to pay back to himself after some time has passed. The receiver will be alerted when that happens, but the sender hopes it will be too late.

The receiver generates a new key pair and gives the public key to the sender shortly before signing. This prevents the sender from preparing a chain of blocks ahead of time by working on it continuously until he is lucky enough to get far enough ahead, then executing the transaction at that moment. Once the transaction is sent, the dishonest sender starts working in secret on a parallel chain containing an alternate version of his transaction.

The recipient waits until the transaction has been added to a block and z blocks have been linked after it. He doesn't know the exact amount of progress the attacker has made, but assuming the honest blocks took the average expected time per block, the attacker's potential progress will be a Poisson distribution with expected value:

$$\lambda = z \frac{q}{p}$$

To get the probability the attacker could still catch up now, we multiply the Poisson density for each amount of progress he could have made by the probability he could catch up from that point:

$$\sum_{k=0}^{\infty} \frac{\lambda^k e^{-\lambda}}{k!} \cdot \begin{cases} (q/p)^{(z-k)} & if\ k \le z \\ 1 & if\ k > z \end{cases}$$

Rearranging to avoid summing the infinite tail of the distribution...

$$1 - \sum_{k=0}^{z} \frac{\lambda^k e^{-\lambda}}{k!} \left(1 - (q/p)^{(z-k)} \right)$$

Converting to C code...

```c
#include <math.h>
double AttackerSuccessProbability(double q, int z)
{
    double p = 1.0 - q;
    double lambda = z * (q / p);
    double sum = 1.0;
    int i, k;
    for (k = 0; k <= z; k++)
    {
        double poisson = exp(-lambda);
        for (i = 1; i <= k; i++)
            poisson *= lambda / i;
        sum -= poisson * (1 - pow(q / p, z - k));
    }
    return sum;
}
```

7

Running some results, we can see the probability drop off exponentially with z.

```
q=0.1
z=0      P=1.0000000
z=1      P=0.2045873
z=2      P=0.0509779
z=3      P=0.0131722
z=4      P=0.0034552
z=5      P=0.0009137
z=6      P=0.0002428
z=7      P=0.0000647
z=8      P=0.0000173
z=9      P=0.0000046
z=10     P=0.0000012

q=0.3
z=0      P=1.0000000
z=5      P=0.1773523
z=10     P=0.0416605
z=15     P=0.0101008
z=20     P=0.0024804
z=25     P=0.0006132
z=30     P=0.0001522
z=35     P=0.0000379
z=40     P=0.0000095
z=45     P=0.0000024
z=50     P=0.0000006
```

Solving for P less than 0.1%...

```
P < 0.001
q=0.10   z=5
q=0.15   z=8
q=0.20   z=11
q=0.25   z=15
q=0.30   z=24
q=0.35   z=41
q=0.40   z=89
q=0.45   z=340
```

12. Conclusion

We have proposed a system for electronic transactions without relying on trust. We started with the usual framework of coins made from digital signatures, which provides strong control of ownership, but is incomplete without a way to prevent double-spending. To solve this, we proposed a peer-to-peer network using proof-of-work to record a public history of transactions that quickly becomes computationally impractical for an attacker to change if honest nodes control a majority of CPU power. The network is robust in its unstructured simplicity. Nodes work all at once with little coordination. They do not need to be identified, since messages are not routed to any particular place and only need to be delivered on a best effort basis. Nodes can leave and rejoin the network at will, accepting the proof-of-work chain as proof of what happened while they were gone. They vote with their CPU power, expressing their acceptance of valid blocks by working on extending them and rejecting invalid blocks by refusing to work on them. Any needed rules and incentives can be enforced with this consensus mechanism.

8

References

[1] W. Dai, "b-money," http://www.weidai.com/bmoney.txt, 1998.

[2] H. Massias, X.S. Avila, and J.-J. Quisquater, "Design of a secure timestamping service with minimal trust requirements," In *20th Symposium on Information Theory in the Benelux*, May 1999.

[3] S. Haber, W.S. Stornetta, "How to time-stamp a digital document," In *Journal of Cryptology*, vol 3, no 2, pages 99-111, 1991.

[4] D. Bayer, S. Haber, W.S. Stornetta, "Improving the efficiency and reliability of digital time-stamping," In *Sequences II: Methods in Communication, Security and Computer Science*, pages 329-334, 1993.

[5] S. Haber, W.S. Stornetta, "Secure names for bit-strings," In *Proceedings of the 4th ACM Conference on Computer and Communications Security*, pages 28-35, April 1997.

[6] A. Back, "Hashcash - a denial of service counter-measure." http://www.hashcash.org/papers/hashcash.pdf, 2002.

[7] R.C. Merkle, "Protocols for public key cryptosystems," In *Proc. 1980 Symposium on Security and Privacy*, IEEE Computer Society, pages 122-133, April 1980.

[8] W. Feller, "An introduction to probability theory and its applications," 1957.

9

About the Author

Daniel Howell is a professor at Liberty University where he teaches anatomy and physiology to premed and nursing students. Prior to joining the faculty at LU, Howell conducted biomedical research at Duke University Medical Center and McGill University.

Howell discovered Bitcoin in 2020 during the COVID lockdowns and quickly fell down the rabbit hole. After spending the next five years exploring this blockchain technology, Howell founded Blue Ridge Bitcoin to serve as a hub for Bitcoin education in Central Virginia. Blue Ridge Bitcoin offers educational and coaching services and holds monthly meetups in Lynchburg.

Daniel Howell is also the author of *Daily Seed: A Bitcoiner's Devotional* published at block height 850,500.

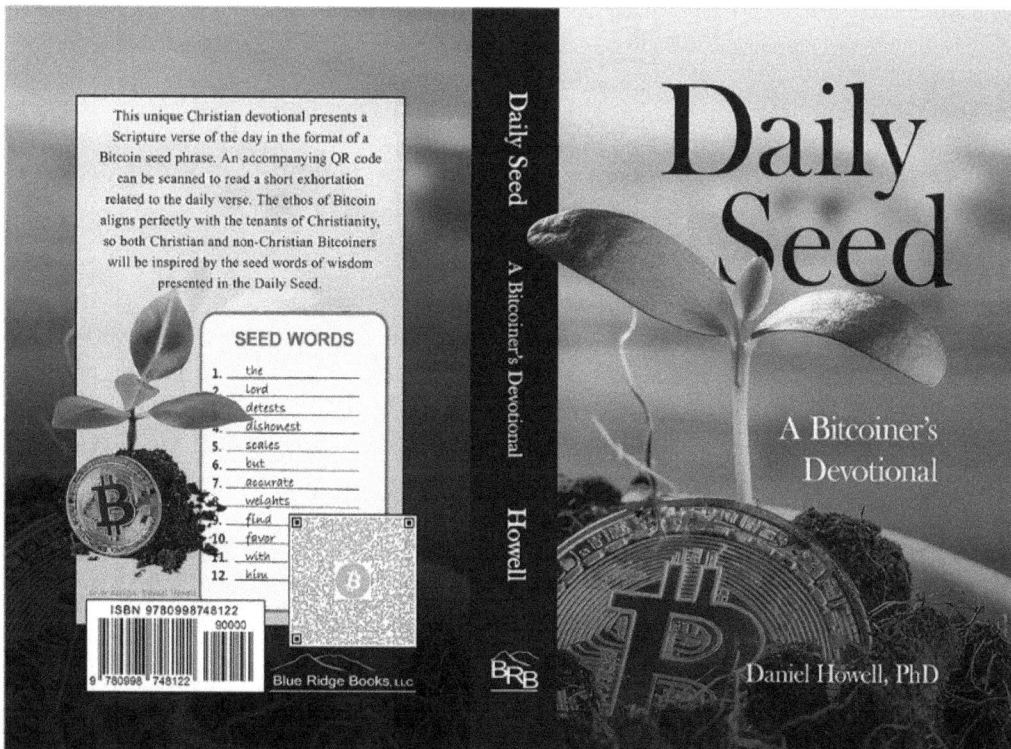

Nostr

If you love decentralized, uncensorable money, you will also enjoy decentralized, uncensorable social media. Nostr is a protocol, not a platform. There is no Mark Zuckerberg or Elon Musk in charge. My favorite app for accessing nostr is Primal. Download the app and give me a follow.

nostr npub for Dr. Daniel Howell

Thank you for reading **Orange Pilled: A Brief Introduction to Bitcoin**.
We would love to get you started on your Bitcoin journey today!

www.BlueRidgeBitcoin.com

BlueRidgeBitcoin.com

Donate sats to BRB at
BlueRidge@mannabitcoin.com

BITCOIN
ACCEPTED HERE

blueridge@mannabitcoin.com

LiGHTNiNG ADDRess